THE REPROGRAPHIC LIBRARY

*Basics
of
Reprography*

Basics
of
Reprography

by

Arthur Tyrrell, F.I.R.T.

Focal Press
London & New York

*Published in co-operation with
the Institute of Reprographic Technology*

PRINTED AND BOUND IN GREAT BRITAIN BY
A. WHEATON & CO., EXETER

CONTENTS

5

materials. Polyester materials. Required properties of diazo compounds. Standard and fast diazo compounds. Other diazo film products. Negative-working diazo systems. Vesicular diazo method. Diazo-metal systems. Diazo compounds in photo-mechanical systems. Positive-working lithographic plates. Image reinforcement by lacquer and baking. Web-coated anodized plates. Bi-metal and tri-metal plates. Etching resists from diazo compounds. Negative-working diazo presensitized plates. Dry lithography.

INTRODUCTION

The total coverage of reprographic processes is wide and to the non-technically trained, intimidating. The purpose of this book is to show principles underlying the processes and not to give detailed instructions on the various techniques. It is intended for use in preparation for the Preliminary Examination of the Institute of Reprographic Technology, but should be useful to others who wish to broaden their knowledge of reprographic methods without having examination aspirations. It is written without any assumptions of previous scientific or technical training; on the contrary, it tries to relate the technicalities of photo-reproduction to everyday experience.

Reprography includes purely mechanical methods as well as 'photo' processes, as will be clear from the syllabus for the Institute's examinations. The material selected for inclusion has been arranged to serve the purposes of the student but should remain readable for others. To see the range of subjects which must be covered we can turn to the syllabus, which expects students at the Preliminary stage to include in their repertoire:

1. Letterpress printing
2. Gravure
3. Offset lithography
4. Stencil and spirit duplicating
5. Document copying using silver, contact and reflex methods
6. Document copying using silver, optical methods
7. Diazo
8. Ferro-gelatine
9. Ferro-prussiate
10. Infra-red
11. Electrophotography

Moreover, the aims of the preliminary course include practical experience of these processes with the exception of numbers 1 and 2, while 3 is restricted to the handling of a small office offset duplicator.

At first sight there is little connection between the printing processes of numbers 1, 2, 3 and 4 and the remainder; but in fact the conventional methods of printing rely increasingly on photo operations in the preparation of the printing surface. Hence reprography includes the group of methods elsewhere known as photomechanical.

The aims of the Institute's Preliminary course also include revision of elementary mathematics, physics, chemistry and English. No attempt is made in this book to provide a revision course in these subjects, but the following parts of the science syllabus of the Preliminary course are to some extent covered:

1. Light; spectrum; infra-red and ultra-violet; inverse square law
2. Nature of sound
3. Principles of operation and efficiency of light sources
4. Reference to the 'benzene ring' in chemistry of carbon compounds

The treatment of these subjects is non-mathematical, and not a substitute for the systematic considerations which should be given, if not at preliminary stage, then in preparation for the Intermediate certificate.

Of the eleven major processes listed, three receive only passing mention in the main text: stencil and spirit duplicating, infra-red, and electrophotography. These are briefly described in an appendix. Other texts in this series will deal more fully with these methods. The remaining eight processes depend on photo-chemical changes and this establishes their underlying similarities. Thus the present text revolves around the relationship between 'light' and chemical change.

By and large the book falls into three sections. After looking at fundamental concepts there is a central section dealing with the main groups of photochemical processes; and finally there is consideration of the process as a functioning whole. Throughout, it is assumed that the reader is a practising reprographic operator; so the book does not attempt to teach the details of 'how it is done', but to explain principles and show these, as it were, in action.

ACKNOWLEDGEMENTS

It will be evident that while this text is in the main concerned with matters of principle underlying the photo reproduction processes, it nevertheless contains a considerable amount of detailed information. I have tried to ensure accuracy in respect of such detail and I gratefully acknowledge my dependence upon many sources in so doing. Apart from numerous personal contacts, I have of necessity drawn upon material published in journals and text-books and the advice and correction of colleagues. I am grateful to these, and especially to the following:

Mr A. Kobler, for assistance with diagrams on pages 44 and 54; Mr G. W. Smith for preparing the photomicrographs and Ozalid Company Ltd for permission to use them; Mr E. Snell for the technique described in page 366 and his encouragement on the use of diazo intermediates in connection with photo-typesetting.

I am especially glad that the manuscript has at two stages received the detailed, kindly and always constructive criticisms of Dr. D. A. Spencer whose name has been prominent in the photographic field for over forty years. In consequence my account is the more rounded and better balanced than it would otherwise have been.

CHAPTER 1

VISION, LIGHT AND COLOUR

Photography is more familiar than reprography; for the products of photography reach us from the cinema and television screens, newspapers and advertisement hoardings as well as the family album or slide collection. In a less obvious fashion reprography also touches the daily lives of most people. Apart from the forms of engineering which depend upon plans and blueprints, there is today a constant stream of paper-work for the ordinary person: forms, circulars, invoices, statements and final notices. These, like the computer, steadily invade the everyday scene and a growing amount of reprographic work reaches into the home through the letter-box.

This array of material has of necessity one thing in common: it can serve its intended purpose only by being seen and conveys its information solely through being visible. Reprographic efficacy is always judged by the criterion of accuracy in transmitting information. Just as photographs normally require to be crisply defined rather than fuzzy blurs (photography as an 'art form' sometimes adopts a different attitude); so drawings must give accurate information to the engineer and bills must show clearly how much we owe.

Nature of human vision

Before considering the processes of photoreproduction, we shall need to obtain some insight into the process of seeing by which we become aware of a visual message. The familiarity of vision is the biggest obstacle to understanding it, though here is no case for familiarity breeding contempt. We shall touch upon the delicate apparatus employed in the visual process, but even more important than a knowledge of the mechanics involved, is an *awareness* of the use to which we each put the information acquired through the eyes.

Almost without exception animal life, including fish, insects and man, has relied upon seeing for survival; among men there is universal sympathy for those who do not possess this facility, as well as a high

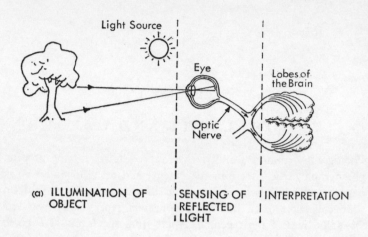

(a) ILLUMINATION OF OBJECT | SENSING OF REFLECTED LIGHT | INTERPRETATION

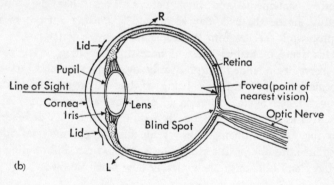

(b)

Light and the eye alone are not sufficient for the process of seeing: *a*, the relationship between light, the eye and the brain. *b*, horizontal cross-section of the human eye (see from above).

degree of respect for the person who overcomes the deficiency. Even so, the majority of people find the process of seeing so natural that they rarely think of its limitations or the ease with which its functioning can be influenced or damaged. 'Take care of your eyes' says the optician's poster, but we subject them without thinking to all kinds of conditions which interfere with or threaten their healthy functioning. In the extreme, some people are submitting their aural and visual senses to such a battering that it cannot surprise us if the whole system revolts. Fortunately there is a growing awareness of the importance of proper lighting conditions for various tasks, for general illumination and road lighting, and special situations such as night driving or television viewing. We all should heed the advice of experts in these fields.

Clearly, sight is connected with light; for in the absence of light vision ceases, while at low levels of illumination we cannot do 'fine' work. In due course we shall look at the nature of light, but for the moment we recognize our dependence upon it: no light – no sight. Each new day brings a welcome return of the conditions for seeing, and this recurrent cycle gives to our existence its most fundamental pattern.

To risk stating the obvious, seeing also involves the eyes; even in sunlight to close the eyes or to be blindfolded is to take sight away. But light and eyes alone are not sufficient for seeing. When attention wanders, we may look at things without seeing them or read a page without recollection of its contents. Call to mind the conflicting evidence of eye-witnesses. They have all 'seen' the same incident, yet quite sincerely fail to agree as to what took place.

Thus, vision is not merely the conveyance of information from the eyes to the brain. This information has to be interpreted before anything is seen to exist. We shall not speculate as to the nature of this interpretation process; it is sufficient for our purposes that we recognize its existence. It is for this reason that 'seeing' is not necessarily 'believing'. The simple process of seeing is evidently quite complex. The eye has to be stimulated by light in order to generate its messages; these must be conveyed to the brain by a nervous system; finally messages received in the brain have to be interpreted. Additionally, there is feed-back from the brain to the eye, so that even this picture of events is over-simplified.

Factors influencing vision

Consider now some of the things which may influence the seeing process. Apart from the presence of light – and even the proverbial cat cannot see in total darkness – there are different qualities of light. The same scene in early morning, at midday, and under artificial lighting gives quite wide differences in its visual impression. Colour is notoriously affected by the quality of light: few people rely on colour judgment under any lighting other than daylight (though this in itself is not of constant quality).

Apart from differences in lighting, an object viewed at the same time by two people may appear different to each. In other words, their eyes have a difference of *response* to the stimulus provided. Even one person may find distinct differences between the impressions gained through his left eye and those through his right eye.

When it comes to the interpretation of the data supplied to the brain from the eye, differences arise when one person has been trained while another is undisciplined or unconditioned by training. If we add to these factors the possibilities of damage to the eye or brain mechanisms, or the effect of ill-health (and which of us is continually 100 per cent fit?) we begin to perceive the total complexities of the simple everyday process of seeing.

In studying reprographic processes we must try to separate out those effects which are real or objective, and those which are connected with the viewer, or subjective. To this end we learn to control and measure light, to recognize various conditions in the normal person's eye, and to be aware of the degree of training which an observer's brain may have received (for example, an artist's or decorator's or dressmaker's appreciation of colour). We may also find that one person's brain has been trained in accurate observation, or may be completely undisciplined, or may be strongly biased towards 'seeing' what it wants to find (in rapid reading, the brain can supply words which the eye has not looked at, through long practice and by deduction from the key words).

Scientifically speaking, the eye is an extremely delicate and sensitive instrument; that is, it can detect or differentiate between small changes in the form and colour of an object. In matters of colour matching, it is very sensitive to small differences, though modern colour measurement surpasses the eye in this respect and is of course independent of the brain's possible interference. Even so, apparatus for the measurement of colour is complex and easily upset; it is best thought of as supplementing direct visual comparisons. 'Remembering' a colour is notoriously unreliable; measurement by instruments is preferable where observations take place over a period of time.

Construction of the eye

Essentially the eye consists of a *lens* which receives light reflected from an illuminated object, and transmits this light into the eye so that an *image* of the object is formed on its back surface. This surface is sensitive to light; it responds to the pattern of the image and generates the information, or signals, sent to the brain. The lens is provided with an iris or diaphragm which can adjust itself to increase or decrease the amount of light admitted; the lens can also alter its shape so that objects can be focused at different distances (i.e., that in each case a sharply defined image may be formed on the back surface). The *cornea* protects the lens and iris from damage.

The back surface, or *retina*, comprises several layers; in the retina are sensitive nerve endings of two types: so-called *rods* and *cones*. The rods respond to dim light, but are not sensitive to colour. Therefore at dusk there is less ability to see colour, and objects tend to appear in tones of grey. Definition is also reduced. In stronger light, the cones come into use. These provide normal colour vision and definition. It is usual to speak of the eye as dark-adapted or light-adapted corresponding to the use of rods or cones respectively.

The light-sensitivity of the rods and cones is due to pigments of which the best known is rhodopsin or visual purple. This is the rod pigment, which is coloured in the absence of light and becomes colourless by exposure to light. In undergoing this change the rod generates signals which are conveyed by the optic nerve to the brain. The cones function in a similar way, but three different substances are involved, giving rise to the sensations which we describe as red, blue and green colours, or admixtures of these. Equal stimulation of all three cone-pigments gives rise to the sensation of white.

In referring to the pigments of the retina as giving *sensitivity* to light, we mean that these substances can undergo changes in structure. The change reverses in the dark and thus sensitivity is restored for further use. It is this change in structure which gives rise to the signal sent to the brain. In technical language, work is done in generating the signal. It is a universal concept that work can be done only when energy is available and used. This is explained in greater detail on page 20. For the moment, the question is: where does the energy which changes the visual pigments come from? It is, in fact, provided by the light itself, which is collectd by the lens and passed on to the retina. The light is *absorbed* in the retina and its energy initiates the changes of which the brain is made aware. Thus we can see why light is necessary to vision – it provides the energy on which the visual process depends, and must be absorbed to make its energy available.

How the eye functions

The eye is an effective instrument over a very wide range of illumination levels. Compare for instance a sunlit beach or snow scene

17

and a moonless night with only starlight available. In order to deal with this wide range of conditions, with its corresponding range of energy levels thrust upon the retina's system, there has to be *adaptation* of the system. This adjustment of sensitivity is a relatively slow process. It is the function of the iris to give the retina time to make this adjustment; the iris can respond almost immediately to large changes in illumination level, thus avoiding the situation where the retina is receiving too much or too little light-energy for its particular state of adaptation at the time.

Alteration to the size of the iris requires muscular effort within the eye, which may involve fatigue. Also, if the illumination level remains high so that the iris must remain at a minimum size for a long period, fatigue arises and causes discomfort. The retina can adjust itself to a brightness range of about 1:50,000; as the iris can superimpose an additional range of 1:20, the eye's total sensitivity range is some 1:1,000,000.

There are many other interesting facts about the eye and its functioning. Among those of importance to reprography are:

Visual acuity: A small area of the retina on the axis of the lens is better equipped to discern detail than the remainder. This is the *fovea*, containing cones only, each with 'single line' connection to the brain. Elsewhere, groups of rods or cones are attached to a single nerve fibre. Thus for critical examination in detail the eye must look directly at the object.

Eye movement: The eyes are constantly making small movements, which cause the position of the image on the retina to change. This prevents some areas of the retina becoming so saturated with light that they would cease to function.

Accommodation: Adjustment to *distance* by focusing the image on the retina through muscular control of the lens curvature. Any object nearer than five feet requires this effort. There is a 'near point' for each individual, dependent on age (which reduces the flexibility of the lens).

Persistence of vision: The effects of light in the retina, and therefore the perception of an object, persist for a fraction of a second *after* light stimulus ceases. "After images" usually give sensations of an opposite (complementary) colour to the original image. Persistence of vision makes possible the moving picture of the cinema screen; the eye is not aware of the 16 or 24 separate frames projected per second.

Colour fatigue: Prolonged observation of a coloured area leads to fatigue, when the brightness of the area apparently is reduced.

18

These effects occur unconsciously: they take place without any deliberate intention on our part. Thus it is important to be aware of them. It may be necessary to have a change of occupation rather than attempt a long, unbroken spell of critical work; or to ensure that opinions are not falsified by changes in illumination (as between daylight at different periods of the day, or between different types of artificial lighting).

Thus the process of vision involves a subtle relationship between light and the coloured pigments of the retina, whereby energy provided by the light is used to set in motion signals to the brain. This is referred to as a photochemical situation, and the consideration of photographic processes will be found to involve very similar relationships between light-energy and chemical change. For this reason, the IRT syllabus requires reprographic technicians in their preliminary course to revise their knowledge of physics (light, heat, optics, sound and electricity) and chemistry. Some knowledge of the fundamental nature of light is essential and the next section deals with this, more especially for those who have little or no physics from their schooldays, to revise.

Light as a form of energy

As with seeing, the first difficulty in discussing light is that the subject is so commonplace. Constant familiarity with the basic day and night rhythm of the solar system means that we rarely stop to think about it; we accept the ability to see even more readily than we accept the air we breathe.

We have already encountered the idea that light, essential to seeing, provides the energy that is required for the retinal changes. In fact, what we call light is a form of energy. To merely change one word for another does not leave us any wiser, but we can relate everyday experience to the concept of energy in the sense of it being *the means whereby work is done*. It may clarify the meaning to consider the characteristic of an energetic person which distinguishes him or her from the lay-about: it is the inclination to take action. In much the same way, light is energetic, i.e. inclined to bring about changes which we can recognize as work done. Photoreproduction involves study of the work carried out by the energy form loosely referred to as 'light'.

Energy exists in many forms but all to some extent follow the same patterns of behaviour. For our purposes, the total energy released into

19

the solar system remains constant; but one form of energy can be converted into another and vice versa. More familiar forms of energy are, for instance, mechanical and electrical. Consider the grandfather clock as a mechanism: it functions by having a weight which is pulled by hand into a raised position. As the weight descends its energy is used to drive the hands of the clock. The energy necessary to do this work is given to the weight when it is raised; it is progressively given up as the weight descends, being transmitted to the clock hands by a chain of gears. When the weight reaches the lower limit of its travel, the clock stops unless the weight is again pulled up by hand. Thus work is done in raising the weight, giving it the ability or potential to do work. Work is done by the weight as it descends, its *potential energy* then being transmitted to the moving parts (some is required to overcome the frictional resistance of the gears, and of their spindles in their bearings).

Kinetic energy

A different kind of energy is seen in the head of a hammer used for driving a nail into wood. For the moment assume that the hammer is in use on a horizontal plane – its potential energy then remains unchanged. The muscles of the user impart vigorous movement to the hammer until the head meets the nail. While moving the head possesses *kinetic energy*, which is transmitted to the nail on impact. The 'harder' we hit, the more movement of the nail into the wood. We hit harder by moving the hammer head faster. Thus the weight and speed of the head represents the amount of energy which can be put to use, the movement of the nail against the resistance offered represents the work done.

Transmission of energy

If instead of a nail we hit a cold chisel against a brick wall, the energy of the hammer is transmitted fairly efficiently along the length of the chisel until at the cutting edge it is absorbed by the brick. Should there by a rubber pad on the chisel, the kinetic energy of the hammer is absorbed partially by the rubber and less reaches the cutting edge, so that less work is done for each hammer blow. The energy absorbed by the rubber is lost so far as useful results are concerned, but does not disappear. Most of it could be found in the rubber or chisel or the surroundings, as *heat*. In other words, these

20

areas undergo a rise in temperature. Thus heat can be seen as a condition involving a higher energy state. This becomes important when considering light sources later on.

Conversion of energy into different forms

Similar situations arise in many familiar forms. In the internal combustion engine, the potential energy which the petrol possesses is converted in the explosion process to movement of the piston. In electrical apparatus, heat or motion is obtained by consumption of the energy possessed by electricity; frequently more work is easily seen to be done when the voltage is increased.

Conservation of energy

These examples bring out some of the characteristics of energy in familiar situations which apply equally to light. First, energy can be measured; it can exist in different forms, one of which can be converted into another; it can be transmitted, with varying efficiency or losses. These losses may mean that only part of the energy available has been utilised in doing the work in question; but not that some of the energy has disappeared. The lost energy has leaked away as, e.g. frictional heat; but it can all be accounted for. If we were able to measure all the leakages and the work achieved, these would total *exactly* the energy given up by the original source.

Light energy as radiation

While these examples are relevant to the thought of light having energy by which it can do work, they are in one important respect quite inadequate. For light can do its work at a distance; there is no need for actual contact between the source of light and the object which is to be affected. Because of its ability to act at a distance, light energy is described as *radiant* energy. We are now able to discard the term *light* and substitute for it the term radiation.

Radiation is itself a very comprehensive term which must be more closely defined. But it is more precise and therefore a more useful term than light, which is necessarily linked with *vision*; especially as the reproduction process may operate by radiation which is invisible – that is, to which the eye is not sensitive and to which it makes no response.

Radiation – types and properties

As yet we do not have a complete knowledge or understanding of the ultimate nature of radiant energy as involved in vision or in photo-reproduction. What we can do is give fairly accurate pictures of some aspects of it; but these remain partial and hypothetical explanations of its behaviour. As with all hypotheses, an explanation has to be modified until it fits all the known facts. No single theory explains *all* the facts so far known about radiation, whether visible or invisible. But over a long period of time, the theories in question have been found reliable for certain purposes, so much so that we tend to accept them as if they were the whole truth. They remain explanations, but explanations which if accepted within a certain context allow us to proceed and to discover more.

Wave-motion concept: The first such theory to consider is that these radiations are vibrations, or as we prefer to say wave motions. As a stepping-stone to understanding this idea, we can turn to the more familiar field of sound and music. Common experience tells us that sound is connected with vibration. The note of a drum, a bell, a tuning fork or a stretched wire, is obviously connected with the vibration of the article in each case. We recognize a certain quality of the sound as the note of the object. The bell or tuning fork has a fixed note, while that of the drum or wire can be altered by changing the tension of the skin or wire. Even the unmusical recognize that the note rises under higher tension and falls under lower tension. These terms, rise and fall, are in general use to convey the required idea – but no one imagines that the instrument literally rises or falls; it is the effect on the ear and the *impression* in the brain which changes. The actual change is one of faster or slower vibrations.

To give precise expression to these conditions, the practice is to refer to a note as having a certain number of vibrations a second. To the initiated, the figure denoting number of vibrations a second is just as meaningful as the note itself. What is even more important, is that when a scale of notes is considered as numerical vibrations-per-second or frequencies, it becomes possible to understand many observed facts which otherwise might remain mysteries.

The analogy with sound must not be pressed too far, but is also useful in accepting the idea of energy being radiated. Energy is required to strike the bell or beat the drum. Both can be heard at a distance, owing to transmission of the vibrations through the air. With

22

sound, the transmission occurs as a pressure wave which we actually feel as well as hear when the sound is loud – another way of saying that energy has been transmitted.

Circular wave motion: The concept of radiated energy can be further illustrated by a stone thrown into a pond. The calm surface of the water is disturbed by a circular wave which rapidly increases in diameter, i.e. travels outward until it reaches the bank. The first wave is followed by a series of waves of diminishing size. All have been created by the energy of the stone being transferred to the water, setting the water into motion or vibration which is carried outward or transmitted by the water to the edge of the pond. If we could measure the input energy, and could add up all the energy dissipated by the water and transmitted to the banks, we should find nothing lost and nothing gained.

Should the pond's surface be covered with floating material, the waves caused by the stone would probably not reach the banks, the energy of the waves being absorbed by the floating matter. The waves or vibrations are then said to have been 'damped'.

Differences between light and sound

The note of a musical sound is peculiar to its frequency and nothing else. The energy of the vibrating object is transmitted by air, water or solid materials. But sound waves are quite unlike light waves because they can be transmitted only by a material environment. An electric bell suspended in an evacuated glass container is quite inaudible, though its ringing remains visible. That is, light has passed out through the evacuated space whereas the sound has not. Because of this fundamental difference in behaviour, it is not satisfactory to continue to think in terms of sound waves in trying to understand light.

We must in fact remind ourselves continuously that though it is convenient to speak of light *waves*, this term describes one of the *properties* of light but not light itself. Over the years it has been possible to make progress in many fields by using wave-motion studies to explain observed effects. But to explain effects is not to explain the thing itself; we may be told that wind can blow down trees or remove chimney-pots, and though these statements are seen to be true, we have not explained what wind is. In a similar way, a wave theory of the nature of light has been found necessary and beneficial – but it is not the whole explanation of light, nor is it possible to

23

reach an understanding of reprographic processes solely on a wave theory.

Application of wave theory to light

Some of the properties of light which can be understood from wave theory are:

1. *Reflection:* behaviour at plane and curved 'mirrors' or polished surfaces.
2. *Refraction:* behaviour in passing through translucent materials, either with plane parallel surfaces, or curved surfaces (lenses), and through materials with plane surfaces which are not parallel (prisms).
3. *Diffraction:* behaviour in passing through a 'grating' of regularly spaced alternate transmitting and non-transmitting lines. The colours seen on a beetle's wing-cases are produced by reflection plus diffraction.
4. *Interference:* behaviour involving very thin films, as of the wall of a soap bubble or oil on a wet road surface.
5. *Polarization:* behaviour in passing through certain solutions, or crystals, leading to the method of detecting stresses in glass or plastic parts.

While it is possible to study the subjects listed by thinking of light as a wave motion, we must not visualize light as therefore being a form of wave and nothing more. There are other effects of light for which the wave theory is unable to give an explanation. For instance:

1. The emission of light by a source, e.g. the sun, arc lamps, discharge lamps.
2. The absorption of light by objects within 'reach' of the radiation from a source, e.g. photo-electric effects.
3. The effects which are produced by light which has been absorbed, e.g. photochemical reactions.

Particulate theory of light

To explain and explore these properties of light, it was necessary to devise a 'particle' theory: the assumption that a beam of light contains small particles of energy travelling in the direction of the light 'ray'.

However, the wave and particle theories of the nature of light are each only partial explanations; each theory fits certain areas of observed fact – but neither is entirely satisfactory on its own.

An 'advancing' wave does not require that the water, air, etc., in which it occurs is actually moving in the direction of travel of the wave. Each particle of water merely moves up and down in a regular sequence and in doing so uses its energy to compel the adjacent particle to follow suit. In this way energy is passed along. The particle theory of light supposes that a minute 'piece' of energy actually travels between source and object. Thus radiant energy possesses the properties of a wave motion but travels in space as separate particles or units. These units are definable as quantities of energy and are referred to as quanta (singular: quantum) or, in the case of light, as photons.

The meaning of colour

The colour spectrum of sunlight is familiar as the rainbow, and the action of a glass prism on a beam of white light produces the same range of colours. Early research workers – Sir Isaac Newton is the best known – were curious to study the properties of white light and its constituent colours, and very soon found that the spectrum did not end where the visible effects stopped. They used chemical reactions to record the regions beyond the spectrum, and in so doing contributed to the chemical processes of photoreproduction.

The colours of the spectrum of sunlight always follow the same pattern. They occur in the order violet, indigo, blue, green, yellow, orange, red. There is not a sharp division between one colour and another; they are in bands of different width, each one gradually changing to the next.

Sir John F. W. Herschel found it necessary to use a numerical scale for the different colour bands, taking a 'fixed point' on the scale by noting the position of a narrow band of light which was transmitted by dark blue 'cobalt' glass. This fixed point helps bridge the gap between his observations and the modern use of wavelength.

Today we are better off in considering colour and its related subjects. By knowing that light has the properties of waves, the whole concept of light behaviour relating to colour can be better understood. Instead of relying on the names of the colour (which are merely expressions of the brain's interpretation of the stimulus which the retina has received), it becomes possible to use measurements. Measurement is the basis of all scientific work and reprography has to be seen in the scientific context.

Wavelengths and colours: Once the wave theory of light is accepted as a working hypothesis, experiment can be used to prove how far the theory is correct. Experiment over a long period shows that the wave theory can be accepted and can be a valuable weapon in obtaining further knowledge. In wave theory, the most obvious thing to measure is wavelength. At the moment, no other properties of wave motions are in question and no detailed examination is called for. The length of the wave is not necessarily the distance from crest to crest, it can be measured as the distance between any two points which are in phase. This expression means that the vibrating particles which form the wave are repeatedly performing a cycle of movement. It is the shortest distance between two particles which are passing through *precisely the same* relative position and movement which gives the wavelength.

To measure the wavelength of light or similar radiant energy requires the use of very small units. The standard unit is the nano-metre (nm) which is one millionth of a millimetre (10^{-9}m). By measurement, the clearly visible colours of the rainbow, or spectrum of sunlight, can thus be expressed in the approximate wavelength values shown in the table below.

Approximate wavelengths of colours

Colour	Wavelength in nm
Violet	400–450
Blue	450–480
Blue-green	480–510
Green	510–560
Yellow	560–590
Orange	590–630
Red	630–700

There is no precise point at which we all agree one colour ends and another begins. Therefore to refer to 'green light' or 'orange light' is not precise; whereas to speak of the wavelength 510 to 560nm is completely definitive and no longer *subjective,* i.e. subject to the interpretive intervention of a brain whether trained or untrained. The eye is more sensitive at around the yellow region than it is elsewhere; by more sensitive is meant that its *response* to a given amount of yellow light is greater than the same amount of other wavelength radiations.

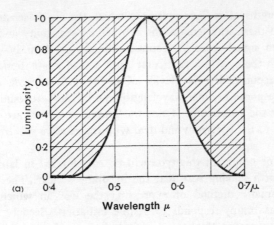

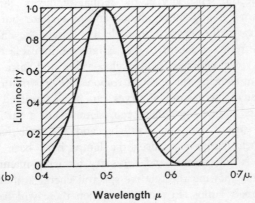

Relative sensitivity of the eye through the visible spectrum: *a*, light-adapted
eye ("cone" vision). *b*, dark-adapted eye ("rod" vision). Based on
"Measurement of Colour" by W. D. Wright (Adam Hilger Ltd,
4th edition, p. 41).

Relative sensitivity graphs

The eye's response to colours falls away to either side of an area of
maximum sensitivity. The position is best expressed graphically and
such a graph is referred to as one of *relative sensitivity*. In this type
of graph, the maximum response is given a value of 100 and the
others are expressed relative to the maximum, i.e. as percentages of
the maximum.

The two graphs showing the sensitivity of the dark-adapted and
light-adapted eye are indicative of the way in which reprographic
situations must be approached. Each curve is based on the range

27

of wavelengths of radiant energy to which the eye is sensitive (below and above these wavelengths the eye has no sensitivity and the brain receives no signal from the retina). The curves then show at which wavelength the eye, in a particular condition of adaptation, shows the greatest response. By the device of giving this response a value of 100, the response to other wavelengths can be measured and expressed *relative to* the maximum. Hence the graphs or curves are of relative sensitivity to the spectrum and deal with the relative *spectral response* of the eye.

Curves or graphs of this type will be encountered in later sections, so it is worth making sure that the meaning is clear. The curves are a diagrammatic method of expressing the way in which the 'eye' (retina plus brain) responds to visible radiation. Because the retina functions by means of photo-chemical changes, the curves must arise from the spectral sensitivity of the retinal pigments. Similar curves can be drawn for any other photochemical system; but in most of these other cases the wavelength scale will extend beyond the limits to which the retina is sensitive, that is into the invisible regions lying beyond the violet and red (hence, ultra-violet and infra-red).

Apart from showing the spectral response or sensitivity of a photo-chemical system, very similar curves can be constructed by suitable measurements, to give a graphical representation of a particular radiation. Thus various types of daylight, and the radiation from arc lamps, filament lamps and discharge lamps have been recorded in this way. It is now normal and necessary for lamp manufacturers to issue diagrams showing the relative spectral energies of the radiation emitted by their lamps. In the next section these will be dealt with in detail. For the moment, it is sufficient to realize that the curves of spectral sensitivity of a system and of spectral energy distribution of a radiation source, are of the same pattern as that of the eye's response, but with the base-scale of wavelength extended beyond the limits of the visible region.

There are other types of spectral diagram, operating on the same principle but rather different in character. They illustrate, for example, the spectral *transmission* of coloured glass or plastic, or the spectral *reflection* of a coloured opaque surface.

To some extent, we are aware of the situations described by these diagrams but without realizing it. For instance, in the darkroom, the yellow light which is safe for bromide paper is unsafe for ortho-chromatic film. The dark red which may be used for the film is quite unsuitable for handling panchromatic material. We are then merely saying that the spectral sensitivity of the three materials differs in that bromide paper is insensitive to yellow light, orthochromatic film is insensitive to dark red light, and panchromatic film is sensitive to most visible light (only very dim green light is usable).

A somewhat different type of spectral diagram to those listed is the *absorption* spectrum of a coloured material. This expresses the same information as the transmission or reflection diagrams, but does so in terms of the relative absorption at each wavelength instead of the transmission or reflectance. This is akin to the photographic terms of transmission and density; the information is the same, but is expressed by the method which is most convenient for the purpose in hand.

Electro-magnetic radiation

Now that we have some familiarity with the use of wavelength measurements in place of colour descriptions in the visible region, we can consider the extension of the wavelength scale in both directions. We soon find that the wavelength band of visible light is a very small portion indeed of the whole. Some of the properties of radiation on the wavelength scale appear to change considerably; for example, X-rays and wireless waves are encountered. All these radiations, including those of light, can be expressed as having both electrical and magnetic properties; they are, so to speak, compounded of the two and therefore the general term for them is *electro-magnetic* radiations. Despite apparent dissimilarities, all electro-magnetic radiation is basically the same, sharing the same behaviour in, for example, reflection and refraction.

In terms of the wave theory of radiation, it is customary to regard the electrical properties of the wave motion as operating on a plane at right angles to that on which the magnetic effects occur. It is not necessary to deal with this subject in greater detail.

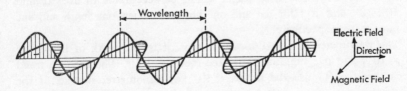

An electromagnetic wave presented as a perspective view of the electric and magnetic fields, which are at right angles to the direction in which the wave "travels".

29

The wavelengths of visible light were quoted above in nanometres. However, the Swedish physicist, A. J. Angstrom, used a unit equal to one ten-millionth of a millimetre which came into general usage as Anstrom units, abbreviated to A.U. or simply A. As this unit is based upon the millimetre, it is of course metric in origin; but the intention to adopt the metric system goes further than this. During the next few years, the basic units in use will be those of the *Système International d'Unites* and many familiar expressions will disappear from ordinary use. The Angstrom unit does not find a place in the international system, but is so widely used that it may be retained for some years and therefore it appears in Table 1:2 for the convenience of those to whom it is familiar.

The complication of changing from the metric system hitherto used in school and elsewhere in this country, is that the primary units were the centimetre, the gram, and the second (hence C.G.S. system). Larger and smaller units were obtained by multiplying or dividing repeatedly by ten; and some of these multiplication or division units came to be known or christened with names that had no systematic meaning. In the S.I. system, the primary units of length, mass and time become the metre, the kilogram and the second ('M.K.S.'). Other basic S.I. units are the ampere (electric current), the kelvin (temperature) and the candela (luminous intensity).

Apart from the change of centimetre, gram, second to metre, kilogram, second, the S.I. requires that larger and smaller units be obtained by multiplication or division, not by ten but by one thousand. In scientific work these multiplications and divisions are expressed as 'powers of ten', thus 10^2 is one hundred, 10^3 is one thousand; in dividing, 10^{-2} is one hundredth, 10^{-3} is one thousandth. The S.I. requires that the basic units are used to generate larger and smaller units only by using powers of ten which are multiples of 3: 10^3, 10^6, 10^9, etc., for successive multiplication by 1000; and 10^{-3}, 10^{-6}, 10^{-9}, 10^{-12} for successive division by 1000. Only during the transition period of C.G.S. to S.I. usage, will it be acceptable to use 10^1 and 10^2, 10^{-1} and 10^{-2} for ten and one hundred times, one tenth and one hundredth respectively.

Unfortunately for the Angstrom unit, it is one tenth of the metre $\times 10^{-9}$, and one hundred times the metre $\times 10^{-12}$. The modern trend is therefore to use the metre $\times 10^{-9}$, or nanometre, instead of the Angstrom.

In presenting a comprehensive picture of radiant energy and the range of electro-magnetic radiation it may be useful to tabulate both

Wavelengths of Electromagnetic Radiation, C.G.S. system (visible light region: 4000 to 7000 A)

Unit: Abbreviation:	metre m	centimetre cm	millimetre mm	micron μ	millimicron mμ	Angstrom A
	1000	10^5				
	100	10^4				
Radio	10	10^3				
Waves	1	10^2	1000			
		10^1	100			
		10^0	10			
		10^{-1}	1	1000		
		10^{-2}		100		
		10^{-3}		10		
Infra-red		10^{-4}		1	1000	10,000
VISIBLE LIGHT		10^{-5}			100	1,000
Ultra-		10^{-6}			10	100
violet		10^{-7}			1	10
		10^{-8}				1
X-rays and		10^{-9}				0.1
Gamma		10^{-10}				0.01
rays		10^{-11}				0.002
		10^{-12}				
		10^{-13}				

Wavelengths of Electromagnetic Radiation, S.I. system (visible light region: 400 to 700 nanometres)

	metre	Multiple or fraction unit	Unit Abbreviation
Radio Waves	$\times 10^3$ $\times 10^0$ $\times 10^{-3}$	kilometre metre millimetre	km m mm
Infra-red	$\times 10^{-6}$	micrometre	μm
VISIBLE LIGHT			
Ultra-violet			
X-rays and Gamma rays	$\times 10^{-9}$ $\times 10^{-12}$ $\times 10^{-15}$	nanometre picometre femtometre	nm pm fm

the older method and the new. In the first table, referred to as C.G.S., the key is in the second column, where every multiple or division by 10 of the centimetre is shown; to the left and right of this column appear the units which had become convenient in dealing with the various types of radiation. The second table shows how the S.I. use of the metre relates to the various radiations and the names to be used for its multiples and fractions.

CHAPTER 2

HOW LIGHT IS PRODUCED

The term light source is frequently applied to any device or apparatus which generates radiant energy, even though the radiation may predominantly be in the invisible regions of the wavelength scale. Very few light sources provide their own energy for this purpose; the majority act as *converters* and require that energy be supplied to them, for example electricity or heat, in order that they can emit radiation. The conversion process is carried out by the molecules of the radiating substance, which is usually a solid or a gas. It is important to have some idea of the molecular state of materials in these different physical forms.

Experience tells us that many substances exist in any of the three states of matter: solid, liquid, or gaseous. Thus water is encountered as solid ice, liquid water or gaseous steam; metals can be melted to liquids and at sufficiently high temperatures will vapourize. Mercury can be frozen to a solid or heated to form a gas, etc. In the liquid state a substance contains more energy that it does as a solid, and in a gas, more energy than when liquid. In these changes of physical state, energy is added or taken away as heat.

Molecular structure of materials

Today we accept that materials are built up from minute particles known as atoms, and that various kinds of atoms combine together to form molecules. Elementary substances form molecules of the same kind of atom, for example iron, lead, zinc, sulphur, oxygen. Compounds comprise molecules of two or more elemental atoms, bound together by forces which may be strong in stable compounds or weak in unstable compounds. In the molecule of a compound, it is not only the number of the various atoms present which decides the properties of the compound, but also the way in which they are combined, or bound together.

In a solid substance, the molecules are in fixed positions relative to each other, although they are in a state of vibration; the higher the temperature the greater the energy associated with the molecules and the higher their rate of vibration. As temperature increases, certain properties of the solid undergo changes; expansion means that the original amount of material occupies a larger volume or, in other words, its density is reduced. This means that the molecules are more free to move. Eventually the solid may melt and in this liquid state the molecules have acquired sufficient energy to overcome to some extent the forces which held them together in the solid state.

The most noticeable change taking place when a solid becomes liquid is the mobility of the liquid substance; in this condition the molecules are much more free to move, though still held fairly close to one another. If temperature continues to rise, the molecules acquire sufficient energy to break away from each other. They spring out of the surface of the liquid and can move with great freedom in the space above the liquid.

In a gas, the individual molecules are much farther apart than in either the liquid or solid state. They move about quite freely, bumping into each other and bouncing off the walls of the vessel (the pressure of a gas in a closed space is due to the combined effect of the molecules colliding with the sides of the container). It is fairly obvious that these free-moving molecules are in a more energetic condition than they were in the solid or liquid states.

For practical purposes light sources are regarded as falling into different types or classes; but an effort should be made to recognize that *all* types of radiation involve the behaviour of molecules under some condition of excitement. Energy is absorbed, gives rise to an excited state, and radiation is emitted. This is consistent with the fact that a light source is creating radiation by a process of energy conversion. It is customary to consider this conversion process under various headings; in practice any one source may operate by more than one conversion method. For a systematic consideration of radiation sources, it is necessary to have a knowledge of:

1. The nature and quantity of the energy absorbed.
2. The type of radiation emitted, i.e. the wavelengths involved and the way in which energy is distributed through the wavelength regions.

The aim must be to convert input energy as efficiently as possible to the desired radiant energy. Much of the progress made in recent years has been in improving the efficiency with which light sources convert energy, rather than in any new principle.

Methods of obtaining radiation

Purely as a convenience in treatment, we can recognize three main mechanisms for obtaining radiation. First, the well-known effects of raising the temperature of a solid substance. This produces a useful amount of visible light, more or less white in colour – that is, stimulating the eye uniformly through the spectrum.

Secondly there are discharge lamps, which rely on the effect of electricity passing through a gas. Examples are the mercury (blue-green) and sodium (orange) lamps of street lighting, and the neon and other tubes of display signs.

The third system relies on *fluorescence*. This is a property of certain solid substances, whereby they absorb radiation of one wavelength region and immediately re-emit the absorbed energy at a different wavelength. In the case of true fluorescence, the emission of the new radiation ceases as soon as radiant energy is no longer absorbed. Sometimes the emission continues for a short time, and is then described as *phosphorescence*.

Categories in practical light sources

Although we shall describe practical light sources under one or other of these three types, it is important to realize that all three are different forms of molecular behaviour. Thus they each involve the absorption by the molecule of energy in one form or another, and the behaviour of the molecules in this excited state. Our categories of convenience are frequently referred to as:

1. *Effects of incandescence.* These arise when solid substances are heated to a sufficiently high temperature, hence thermal radiation.
2. *Atomic radiation,* being typically the result of electrical excitation of the molecules of a gas (as in discharge lamps).
3. *Luminescence,* resulting from absorption of radiation by solids and re-emission at a different wavelength.

Luminescence is usually associated with materials operating at or about room temperature. When the re-emission persists or continues after excitation the term phosphorescence applies; when it ceases as soon as excitation is stopped, the term fluorescence is used.

Bearing in mind that the emission of radiant energy is a molecular process, it is not surprising that the three categories of practical light sources show differences which reflect the molecular state of the substance concerned. In terms of the molecular mechanism of producing radiant energy, the logical classification of sources would seem to be:

1. In cold solid materials: fluorescence and phosphorescence.
2. In hot solid materials: thermal radiation.
3. In gases: atomic radiation.

Thus while incandescent effects may be more familiar in everyday life, we shall consider fluorescence first.

Fluorescence

Examples of fluorescence are often displayed in museums dealing with geology or naturally-occurring crystalline materials (minerals). A cabinet of these is illuminated alternately by ordinary tungsten filament light and ultra-violet or 'invisible' light. Substances which appear merely greyish or white under visible light glow with a variety of colours when illuminated by ultra-violet light. This is because they have the ability to absorb the energy of the ultra-violet radiation and convert it to radiation at longer wavelengths, frequently in the visible region – hence the 'glow in the dark'. This is the principle on which the fluorescent tube works. The same principle exists in the screen of the cathode ray tube of television; the coating on the screen is excited by the electron beam, i.e. it absorbs the energy of the beam and re-emits energy of visible wavelength. In this sense the screen is a light source, and not a screen in the sense of a projection screen for ciné. A third example is the use of intensifying screens in X-ray photography. X-rays which have penetrated the 'soft' parts of the subject are absorbed by the screen and the energy is re-emitted at wavelengths to which the photographic emulsion is more sensitive than to X-rays themselves. In this way the photographic effect is intensified so that shorter exposures are possible.

Fluorescence of the television screen type is not used alone as a means of room lighting. But it is used to very great practical effect to *modify* the emission of gas discharge lamps. This is of immense importance, because it allows an arrangement for efficient conversion of electrical energy to radiant energy, and a further conversion of invisible radiant energy to light.

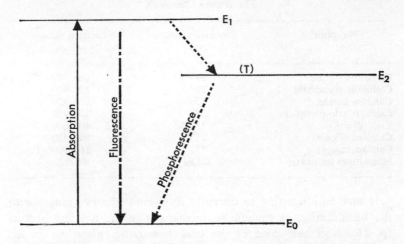

Molecular energy levels and light emission.

The action of fluorescence can be represented by an energy-level diagram, which can illustrate the process taking place without requiring the complex mathematics proper to a study of molecular behaviour. In such a diagram the base line represents the energy content of a single molecule of the fluorescent substance, when it is in its normal or un-excited state (E_0). By absorption of energy, the energy content is raised from this ground state to a new higher level (E_1). The molecule tends to return to its stable ground state by discarding the newly-acquired energy. If it does this instantaneously and completely by emitting radiation, we have fluorescence; if it first drops to some intermediate state (E_2) – a sort of storage of the acquired energy by some internal re-arrangement – it can discard energy over a period of time, eventually returning to the ground state. In this case, the effect is phosphorescence.

In the lamp industry, it is usual to refer to the fluorescent substances used as phosphors. As the only difference between fluorescence and phosphorescence is the *time* or duration of the effect, there need be no confusion.

As the fluorescence/phosphorescence effect is bound up with the energy-states of molecules, it would not be surprising if different substances showed differences in behaviour. This is the case, as may be seen in the table overleaf, adapted from that compiled by Kosar (*Reprographics*, January 1967, p. 9):

37

Phosphor	Sensitivity peak nm	Emitted range nm
Cadmium silicate	240	480–740
Cadmium phosphate	247.5	270–400
Calcium borate	250	520–750
Calcium halo-phosphate	250	350–750
Zinc silicate	253.7	460–640
Calcium silicate	253.7	500–720
Calcium tungstate	272	310–700
Magnesium tungstate	285	360–720

It may be interesting to interpret these emitted wave ranges with the identification of colours by wavelength in the previous section (p. 26). Note that none of the peak sensitivities given are in the visible region, whereas most of the emitted light is. Peak sensitivity may be thought of as the wavelength of maximum absorption. Thus the different molecules can absorb energy of different wavelengths as a property of their internal structure, and re-emit at wavelengths which are typical of the substance. This behaviour is said to be *characteristic* of the substance, in which respect there is a fundamental difference between luminescence, and thermal radiation of heated solids (incandescence).

Incandescence

There are two major differences between radiation from incandescence and that from fluorescence. The first is that *all* solid substances when heated to a given temperature, emit much the same radiation; in other words, the type of radiation is more characteristic of the temperature than of the substances involved. The second is that whereas the phosphors cited above each radiate in a particular wavelength region, all incandescent solids emit radiation over the same wavelengths. This position leads to the use of a scale of temperature as a means of describing or defining the visible radiation from an incandescent solid.

The fact of incandescence is recognized in such common terms as 'red-hot' or 'white-hot' and it is found that as temperature rises the colour of light emitted passes from red through yellow and then to white. A scale of approximate temperatures corresponding to changes in the appearance of heated materials is given opposite.

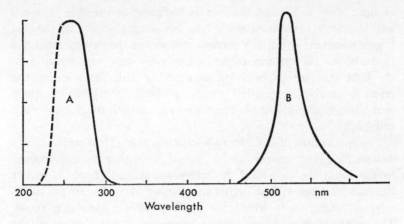

Excitation (A) and emission (B) ranges of wavelength for activated zinc silicate. From Dushman, Gaseous Discharge Lamps, *JSMPE* **30** 79 (1938).

These changes are due to a progressive shift from the red radiation first emitted to include more of the spectrum right up to the blue region.

As a light source, incandescence is concerned with visible light; but long before a solid reaches the temperature at which its radiation

Temperatures of heated bodies

Heated appearance	Temperature
	°C
Incipient red heat	500–550
Dark red heat	650–750
Bright red heat	850–950
Yellowish red heat	1050–1150
Incipient white heat	1250–1350
White heat	1450–1550

(Reproduced from the Handbook of Chemistry and Physics, 41st Edition).

is visible, it is emitting in the extensive infra-red region. There is a tendency to refer to this as radiant heat but we should be careful to avoid this expression. Infra-red radiation is not itself hot. It is produced by a heated substance, and when *absorbed by another material* produces a higher temperature in that material. When we hold our hands in front of an electric iron, they give us the impression of heat. This is because the iron is radiating at invisible infra-red wavelengths; the radiation is not hot, but when absorbed by the skin it generates heat which the nervous system can detect and announce to the brain. The eye sees colour in much the same way, not because the light rays are coloured but because the radiation reaching the retina is capable of stimulating messages which the brain interprets in the language of colour (to which we have all been conditioned since childhood).

The mechanism of solid incandescence is that of molecular effects due to the higher energy level achieved by raising its temperature. In the solid, the individual molecules are packed closely together; although they can vibrate, they cannot move about as can gas molecules travelling freely within the confines of the containing vessel. The radiation from incandescence cannot be 'characteristic' of the heated substance because of this limitation on the molecule's freedom to move. Nevertheless, as more and more energy is packed into these closely confined (or bound) molecules, their vibrational behaviour changes and emission of electro-magnetic waves takes place. The resultant radiation is similar for all types of molecule when raised to the same temperature. We shall see later that when gases are highly compressed there is some restriction on the freedom of movement of the separate molecules – more collisions with each other in a given time. Resemblances between the thermal radiation of solids and the atomic radiation of gases are then found.

Incandescence and colour temperature

The *principle* of incandescence is mainly expressed in the concept of the 'ideal black body' or full radiator. This is a theoretical concept of a solid substance with a surface which completely absorbs all radiation which falls on it (i.e. reflects none, hence, perfectly black). Such a substance would also be a perfect radiator in the sense that its radiation is solely due to its own incandescence and is not confused with reflected radiation from other sources.

The practical results of incandescence can be conveniently and reliably expressed as the radiation of the ideal black body when heated to some definite temperature. The sun is found to show the behaviour of such a black body, and sunlight can be regarded as the result of incandescence at some specific temperature.

As raising or lowering the temperature of an incandescent solid immediately produces predictable changes in its *colour*, we arrive at the use of *colour temperature* as specifying a particular kind of radiation. This temperature is measured on what is called the *absolute* scale of Kelvin units (K), after the physicist Kelvin. The Kelvin scale uses the same size of unit as the Centigrade and Celsius scales, which is the difference between freezing point and the boiling point of pure water at 760mm barometer pressure divided by 100. But the Kelvin scale places its starting-point at what is called absolute zero, corresponding to a temperature of 273 degrees *below* the zero of Celsius. Thus 273K = 0°C.

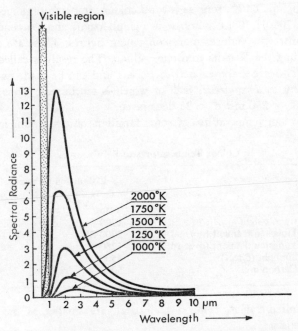

Changes in radiation from a "black body" (full radiator) with temperature. Bauer, *Measurement of Optical Radiations* (Focal Press, 1965).

The colour temperature of a light source is the temperature in kelvins to which the ideal black body would have to be heated to produce light of the same spectral qualities as the source. It is *not* the actual temperature although with say a tungsten filament lamp, the colour temperature rises as the filament temperature increases.

The radiation from an incandescent solid is continuous, i.e. it occurs continuously through a given wavelength region. The radiation from many practical sources may be discontinuous: it may cover

41

the same extremes of wavelength but have no emission at all at certain points. In such a case, the allocation of a colour temperature to the source can be an approximation only; nevertheless it remains a convenient method in practice of specifying the quality of light for a given purpose – notably in colour reproduction processes. Where the light source does in fact function by the incandescence of a solid material, its radiation can be a very close match to that calculated for an ideal black body at some specified temperature – but see the description of the Welsbach mantle and carbon arcs (pages 54 and 65).

Reprographic light sources cover a colour temperature range of some 2000 to 7000K. These numbers are cumbersome and a second system is in use which produces more convenient figures. These are obtained by dividing the K units into one million. The result is called *micro-reciprocal-degrees* abbreviated to *mired*; and can be made even more convenient as *decamireds*, each of which is ten mireds. Thus:

4000K = 250 mired = 25 decamired.

The colour temperatures of some familiar sources are for instance:

Colour Temperatures of light sources

Source	Colour temperature K	Mired
Candle-light	1700	590
Tungsten filament (normal)	2000–3000	500–330
Tungsten filament (over-run)	3250	307
Sunlight (direct)	4500	212
Carbon arc	5000	200

The mired values, being a reciprocal, are smaller as the K unit value increases.

Although colour temperature is a convenient way of expressing the quality of thermal radiation by a single figure, it is not always precise. For reprographic purposes the spectral energy diagram may be necessary. For example, a colour temperature may be assigned to a light source which is a close match to daylight *in the visible* regions, but deficient in the wavelengths required for a reprographic purpose.

Atomic radiation

Gases are employed in closed spaces – to prevent their escape – technically termed envelopes. Sometimes the gas is used at a pressure

42

many times atmospheric in which case the envelope must have the requisite mechanical strength. The envelope must also have high transparency at the wavelengths involved.

The method of producing radiation from gases is to submit them to the influence of a high electrical potential (voltage). Electrodes are introduced through the envelope and are connected to a high voltage supply. This produces a voltage differential between the electrodes and, if conditions are right, a discharge occurs between the electrodes; hence, 'discharge lamps'. The discharge appears as a brightly glowing column between the electrodes which may be referred to as the 'arc' (by reference to the carbon arc lamp which operates in air at atmospheric pressure).

Discharge lamps are necessarily more expensive to make than other types of source, and require electrical gear to produce the required high voltages. Hence they are initially expensive to install, but operate at a high efficiency in terms of radiant output to energy input (and therefore, lower running costs than say incandescent lamps of the same total light output).

Under the conditions inside the discharge lamp, the mechanism of producing radiation is again molecular in character; but in this case, the gas molecules are in a very different state from those of the fluorescent or incandescent solid. They are free to move about within the space enclosed, merely colliding occasionally with each other or with the envelope. Each of the atoms or molecules is therefore free to behave in its most natural way and without interference from its neighbours. This behaviour is intimately related to its structure and the electric field in which it finds itself.

By field is meant the stream of electrons which passes from one electrode to the other under the pressure of the applied voltage. It is the movement of electrons which causes electricity to flow. Thus the atoms or molecules of gas in the discharge lamp are not only free to move about, but are moving through a stream of rapidly moving electrons. Each atom contains one or more electrons revolving round a central nucleus and so it is not surprising that some of the energy of the electric field is transferred to the gaseous atoms which are therefore in an excited state. This state of excitement is usually short-lived: the atom quickly discards its surplus energy, returning to its normal state. The surplus energy may be passed on to another atom, or give rise to heat when the atom collides with the envelope, or be transformed to radiation.

In the case of a simple atom with only a few electrons in its struc-

43

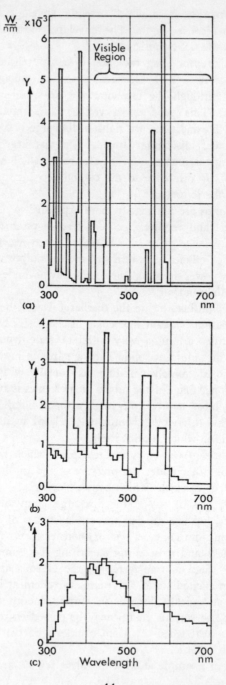

(a)

(b)

(c) Wavelength

ture, the emitted radiation might be almost entirely at one wavelength. More complex atoms offer greater possibilities for successive stages in the energy-reduction process; in this case, the resulting radiation might occur at several wavelengths – one for each stage. Thus the radiation emitted from an electric discharge through a gas is *characteristic* of the atoms or molecules of the gas. On the spectral diagram, these emissions appear as isolated lines at the various wavelengths, as opposed to the broad bands of the fluorescent substance.

Continuous and discontinuous spectra

We have already spoken of incandescent solids as producing *continuous* emission in the sense that radiation occurs continuously over the whole wavelength range involved. The diagram representing this emission is referred to as a continuous spectrum, whereas the similar diagrams relating to fluorescence and gas-discharge radiation are *discontinuous*; discontinuous spectra may show lines or bands. Although these terms are widely used, the distinctions made are somewhat artificial because the one tends to merge into the other. However, they are a working convenience and no harm is done so long as we realize that the differences are not fundamental. An effort should be made to recognize the similarities and underlying principles of light sources in molecular mechanisms, because this is the reason why the photo-chemical changes of reprography are possible.

As examples of typical line spectra, see pages 63 and 44 for sodium (11 electrons in the atom) and mercury (80 electrons) when operating at around atmospheric pressure. When, however, the operating pressure is increased the spectral diagram achieves a continuous form, even though peaks persist at the wavelengths of the original lines.

In the main, lamp manufacturers are always seeking to improve lamps in terms of achieving a better resemblance to daylight. Operation of gas-discharge lamps when a single gas is used usually falls far short of this objective. Although operation at higher pressures is

Spectral energy distribution for high pressure mercury vapour lamps at approximately constant current and increasing pressure, potential gradient and input/cm: *a*, at about atmospheric pressure. *b*, at about 75 atmospheres pressure. *c*, at about 300 atmospheres pressure. The output of radiation energy (Y) is shown at watts per nm, per watt of input. From Beijer, Jacobs and Tol, "The iodide discharge lamp", *Philips Technical Review*, 29, 353–362, and Elenbaas, *The High Pressure Mercury Vapour Discharge*, North-Holland Publishing Company, Amsterdam (1951).

beneficial in terms of imparting to a line spectrum a continuous background, the results with, say, mercury are still far from what is required as a 'white' light. We shall see when considering practical light sources that two main possibilities exist of using the efficiency of discharge lamps for a closer approach to daylight:

1. Combination of fluorescence with mercury vapour discharge.
2. Admixture of other substances with mercury in the discharge.

For various reason these two approaches seem to have advantages over the discharge lamp using the gas xenon, though a glance at the spectral emission curve for this gas shows its close resemblance to daylight. On this account xenon lamps are attractive as illuminants for the camera copy-board especially in colour work. Lamps derived from mercury vapour as above, may be preferable in reprographic operations (see page 213).

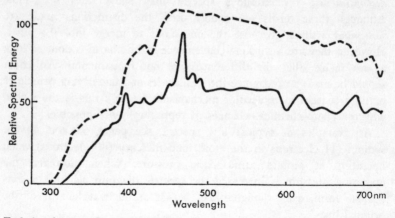

Emission from an Osram XBF high-pressure water-cooled xenon discharge lamp (solid line), compared with "average" daylight (broken line). The xenon discharge has a peak emission in the infra-red region, between 800 and 850nm.

46

CHAPTER 3

PRACTICAL SOURCES OF LIGHT

In considering the principles of light production we have established
the idea of energy conversion and have seen that this is a molecular
process. For practical purposes it must be possible to start or stop
this conversion process as and when required, which is normally
achieved by cutting off the input energy, usually in the form of heat
or electricity.

Efficiency of energy conversion

Different converters behave differently in their speed of response to
starting and stopping the energy input: there may be a warming-up
period or a cooling-down period, or there may be instant and immediate
commencement and cessation of radiation. This, and the efficiency of the
conversion process, are obviously important practical considerations.
Efficiency is to be judged not only in overall conversion of absorbed
energy to radiated energy, but in the efficiency with which radiation
useful to the reprographic process is produced. This depends of
course on the reproduction process involved and the choice of a
light source is therefore compounded of many factors. The spectrum
of an emitted radiation is one key to its suitability for a given
purpose; hence the importance of spectral diagrams.
Apart from light sources which can be brought into use as and when
required – that is, are under our control – the sun is the one great
prime source of energy on which we depend. Our world has derived
all its energy from the sun, and is kept supplied with energy which
radiates from the sun. The molecular processes which we can recognize in our everyday light sources, are also taking place in the sun
but on an incredibly grand scale. Measurements and calculations have
been made but the magnitude of the values is beyond the grasp of
most people; yet our sun is only one small star in the universe.
Nevertheless, because the sun dominates the scene in the visible
result of the reprographic process, even if not involved in the process

47

itself, we need to have some idea of the sun's functioning as a radiation source.

The sun as a light source

We are of course dependent on the astronomer and his many years of patient study for this knowledge. The sun is considered to be a sphere of gases, approaching one million miles in diameter. Despite its gaseous nature, the centre is so dense that its behaviour as a radiation source is more nearly that of a heated solid than that of a gas. We are told that 99 per cent of the total mass of material is concentrated within 0.6 per cent of its radius. Thus the centre of the sun is extremely dense and extremely hot – a temperature of millions of kelvins.

These conditions are unlike anything produced on earth and the material at the sun's centre is unlike anything we normally meet. This material must exist as the basic units of matter (neutrons, protons, electrons) and not as the associated units we could recognize as chemical elements (iron, lead, carbon, etc.). It seems to be generally agreed that at the centre of the sun, the conditions are such that matter is being converted into radiation. Energy is continuously interchanging between particles, giving rise to radiation; and whereas the particles are trapped at the centre by gravitational forces, the radiation energy can eventually find its way outwards and in time escape into space.

The photosphere

This very dense centre is surrounded by layers of gas which are cooler as they are farther from the centre. Sunlight comes from these outer layers, and not from the centre. If we look at the sun, we look into and through these layers but cannot 'see' beyond a certain point. This is the relatively thin layer in which visible radiation escapes, hence it is known as the photosphere. Whatever the nature of the processes taking place in the centre, the gases of the photosphere radiate as if they were an incandescent solid, i.e. the radiation shows a continuous spectrum, not a line or band spectrum as we would expect from a gas at atmospheric pressure.

As the radiation progresses outwards from the photosphere, it must pass through a gas layer some 5000 miles thick. In the outer layer, much cooler than the central regions, the basic units of matter have grouped together to form atoms and molecules which we can recognize by their action on the escaping radiation. For these atoms and molecules, being better able to behave in their characteristic fashion, absorb from the radiation those wavelengths which match their own natural vibration frequency (in much the same way that a violin string vibrates when sound waves of correct vibrational frequency reach it from some other source).

Fraunhofer lines

The net result is that the atoms and molecules present in the outer layers of gas *extract* energy at certain wavelengths from the continuous radiation of the photosphere. Thus the light which finally reaches us on earth is of a continuous nature but has had certain wavelengths extracted from it. When sunlight is analysed by a spectroscope (an apparatus for arranging wavelengths in an orderly sequence) the missing wavelengths appear as dark lines on a continuous background. In the ordinary way we are never aware of these missing parts of sunlight because there are plenty of residual wavelengths to operate the eye's system of colour vision.

The lines are in fact the *absorption spectra* of the substances present in the sun's outer gaseous envelope. Careful examination of these lines, and comparison with line spectra produced from known materials, has made possible the identification of many familiar materials. The first person to discover these dark lines in the sun's spectrum was the English physicist W. H. Wollaston (1802), but it was the German Joseph von Fraunhofer who examined them in detail, publishing his observations in 1814–1815.

How the radiation is produced

Thus the picture we have of the sun as a light source is of a centre in which nuclear transformations generate energy which in due course escapes outwards as radiation; a layer of gas called the photosphere, in which the radiation takes on most of its final character and will no longer be engaged in the absorption-regeneration processes of the centre; and outer layers of cooler gas which extract certain wavelengths of the photosphere's radiation, thereby identifying the substances present in these outer layers. It has been possible to recognize 60 or more elements in this way, and some 18 compounds.

The energy which the sun radiates involves a steady loss in mass, because the sun's process is to convert mass into radiation which passes into space: only a very small portion indeed reaches the earth. Contrary to the thinking of 40 years ago – of a solar system which would 'run down' as the sun *cooled* – we are told today that there is little evidence of it either heating or cooling. This must mean that the energy generated at the centre is balanced by the energy lost by radiation from its surface.

Effect of scatter and the blue sky

Radiation which finally leaves the sun's outermost layers travels through space without further spectral change; but on entry to the earth's atmosphere it encounters dust particles which give rise to scatter. In terms of colour temperature, noon daylight entering the earth's atmosphere at the perpendicular (from directly overhead) can be assigned a value of 6000K. But indirect sunlight from the sky has been enriched in blue light by scatter.

The effect of scatter on a beam of white light is to divert the shorter wavelengths at right angles to the path of the beam. Thus the beam is deprived of blue light, giving it an 'excess' of yellow and red. Scatter is pronounced on foggy days, when the sun is seen as strongly red in colour. Car headlamps approaching through mist are reddish, while an observer on the pavement (seeing from a position at right angles to its direction), is aware of the blueish scatter. Yellow fog-lamps – yellow is white light minus blue – are effective because there is no blue to be scattered. A much higher colour temperature is assigned to north skylight than to noon sunlight – which is another way of saying that the sky appears blue!

For critical purposes of viewing colour – such as painting a picture or matching dyed fabrics – north sky light is used, being more reliably constant and acceptable in spectral composition than any other 'daylight' source. These differences of spectral composition of natural

Colour temperature of various daylights

Type of daylight	Colour temperature K
Direct sunlight	4500
Average daylight	5500–6000
Direct sun + blue sky	6500–7000
Blue skylight	8000–20000

light are shown by diagrams or curves; note that normally each curve is drawn as relative energy, assigning to each wavelength an energy value which is a percentage of the maximum for that source. Thus while both curves reach the same maximum of 100 per cent in the diagrams, this does not mean that each source has *equal maximum*

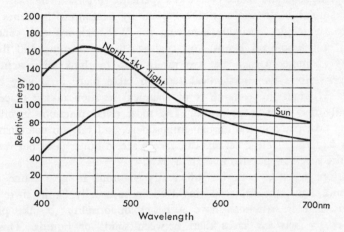

Spectral energy distribution curves for sunlight and north-skylight. From "The Measurement of Colour", W. D. Wright, p. 6 (Hilger & Watts, 3rd edition, 1964).

energy. In absolute terms, direct sunlight might supply many times as much energy at say 400nm than north skylight at the same wavelength, even though on the diagrams of relative energy it may appear to be the reverse. This is an important point to remember in comparing any light sources.

Apart from the dark Fraunhofer lines, the sun's spectrum is continuous; but the curves show that energy is not *equal* at all wavelengths. Its maximum energy is at about 500nm, which more or less coincides with the maximum sensitivity of the human eye. This is not surprising, because the eye has evolved from more rudimentary organs whose original function was to differentiate between dark and light at a time when the sun was the only provider of light. Thus the very highly evolved eye has retained maximum response to the dominant wavelengths of sunlight.

Thermal radiation in artificial sources

The sun's radiation is essentially continuous, as is typical of a heated solid. It is therefore natural to proceed to consider practical sources which also depend on the incandescence effect. There are two methods of heating to produce incandescence: by flame, and by passage of electric current. The review of practical light sources will include some which may be of little or no importance in modern reprography, but which exemplify the principles set out, make the survey more comprehensive and relate everyday experience to these principles.

Examples of incandescence of solids by heating with flames include: limelight, gaslight (involving the Welsbach mantle), candle-light and simple oil lamps, and coal-gas flame. Burning and flame are so commonly met that we rarely stop to think what actually happens. In a flame, chemical reaction takes place at a very high rate. A definition of flame is of chemical reaction taking place so rapidly that light is produced. The most usual reactions involving flame take place in the normal atmosphere, and rely on combination with oxygen. Flame reactions can take place in other gases but would require special apparatus to demonstrate.

Before reactions can be sufficiently vigorous to give flame and produce light, the materials involved must usually be in their gas form; that is, reactions have a better opportunity to take place vigorously between gases than between solids or liquids. This is understandable in terms of their molecular state as previously described. To sustain flame it is necessary to have a continuous supply of the gases involved; hence the forced draught to encourage a boiler fire.

Apart from atmospheric oxygen, in which the combustion is to take place, a supply of combustible material(s) is required. When these are solids like paper, coal and wood; or liquids like petrol or benzene, they must first be raised in temperature until flammable gases are given off. Certain materials are more flammable than others because they produce combustible gases at a lower temperature. Hence paper to light the wood, and wood to light the coal. Coal burns when it is heated sufficiently to give off flammable gases.

Types of flame and effects in flames

Having produced flame, light will be emitted from any solid material heated by the flame to a temperature of about 1200K. Simple flames of the candle, oil or gas type produce light by reason of *incomplete combustion* and the presence in the flame of unburnt carbon. The flame is therefore 'sooty' as can very easily be shown. The light produced comes from the incandescence of carbon particles in the flame; it is yellow in colour, maximum at about 590nm and of continuous spectrum. Carefully examined, the flame is seen to comprize three zones:

1. Inner zone: volatilized but unburnt hydrocarbons from the wick; non-luminous.
2. Central zone: of active but incomplete combustion and therefore luminous.
3. Outer zone: combustion continuing with excess air; non-luminous on account of lower temperature due to cooling.

In terms of energy conversion in simple flames, the 'fuel' – whether wax, oil or gas – is burnt, the combustion process raises the temperature and incandescence follows. Much of the potential energy of the fuel re-appears as heat and is lost as far as production of light is concerned. The advantage of candle and oil lamps is that they are self-contained, that is, they do not rely on any external source for the input energy. It should be noted that the wax of the modern commercial candle is chemically related to the paraffin of the oil lamp, but is of higher melting point. Both are, like coal gas or acetylene, hydrocarbons; most flame illuminants rely on carbon compounds and function by the incandescence of carbon particles.

In these simple flames, atmospheric oxygen supports the combustion, and if the oxygen supply is cut off flame is reduced and extinguished. With the sophistication of a piped supply of gas it becomes possible to supply fuel and oxygen at a constant proportion rather than rely on replenishment of atmosphere by natural convection currents, etc. As a first step in this direction we have the Bunsen burner, which is designed as a better source of heat rather than light – better, in the sense of achieving more complete combustion and higher temperatures. The hotter flame of the Bunsen burner can therefore be used to bring some solid material to incandescence; this is the principle of the gas lighting invented by Welsbach and requiring the use of his 'gas mantle' to provide the radiation.

Better still, both the gas and the air or oxygen required for combustion can be piped under pressure to the burner, and the supply adjusted to give an exact balance for complete combustion. Again, this very hot flame can be used to bring a material to incandescence, and is the principle of the limelight in which the flame is played on to the surface of a rotatable cylinder of lime. The Welsbach mantle provides a large-area source suitable for room and street lighting; the limelight gave a compact source of high intensity, used with reflectors and lenses to project a narrow beam over a distance. This was the 'spot' of stage lighting, hence the expression 'in the lime-light'.

To complete the sequence from candle flame to Bunsen burner and gas/compressed air of limelight, there is the acetylene flame which is not only intrinsically hotter than the coal-gas flame, but used with an oxygen supply gives the oxy-acetylene flame of welding and metal-cutting use.

Principle of Welsbach lamp

As with the carbon particles of the candle- and oil-flame, the Welsbach mantle emits radiation of continuous spectrum, but at a higher temperature and giving a 'whiter' light. The Welsbach burner is simply a Bunsen burner with careful regulation of gas and air supply. The mantle is a fragile structure of 99 per cent thorium oxide and one per cent cerium oxide. The mantles are made by forming cotton fabric in the required shape, impregnating it with a mixture of thorium and cerium nitrates, and drying. The cotton is then burned away at a temperature sufficiently high to leave the mixed oxides as a 'skeleton' of the fabric. For safety in transport the mantle is treated with collodion (see page 94), which quickly ignites and burns away completely when the mantle is first put into use. The radiation from the mantle, while continuous, is said to be *selective* because the infra-red radiation is proportionately less than that of the black body at the same temperature.

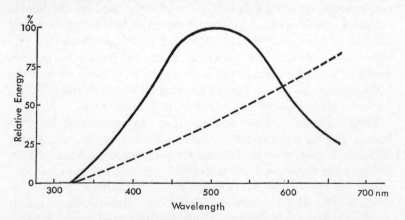

Selective radiation emitted by Welsbach mantle (solid line) at 1800K in comparison with black body radiation (broken line) at the same temperature.

Incandescence of metal oxides

In the hydrocarbon flame, light comes from incandescence of carbon particles of which there is an excess, this eventually giving smoke or soot. In the Welsbach system and in the limelight, the incandescence is of the oxides of metals. These oxides are very stable substances and do not decompose under these conditions. Other metal oxides can also be brought to incandescence and have found use in photography.

There is the familiar magnesium ribbon of the school laboratory. The metal is in the form of a narrow strip and very thin – perhaps 0.001in. When heated to a sufficiently high temperature in air (above 650°C, the melting-point) magnesium burns and gives off an intensely bright, white light. The process is difficult to start – a match flame is not usually hot enough – but once started will continue without any trouble. Clouds of white 'smoke' are emitted in the form of magnesium oxide. It is the incandescence of the magnesium oxide as it is formed which produces light: that is, the burning or 'oxidation' of the metal provides the high temperature and the incandescent oxide provides the light. This has been utilised in photographic flash powder which has practically disappeared from the scene today, but provided much amusement as well as light in the earlier years of this century.

Flash powder and flash bulbs

The aim of a flash powder was to produce very rapid – 'instantaneous' if possible – total oxidation of a measured quantity of magnesium powder. The quantity depended on the size of subject and its distance from the camera. To do this required that a substance be intimately mixed with the metal powder, to provide quickly all the oxygen required for combustion of the metal. Combustion had to be rapid so that exposure was complete before the sitters had time to move! A suitable mixture could burn in about one hundredth of a second, but the photographically useful radiation is greater if combustion is delayed and is best at about one twenty-fifth of a second. The more effective mixtures were unfortunately also the more explosive. Aluminium could be used if extremely fine; other metals could be included such as lithium and thorium. 'Flash' could be prolonged for night scenes by mixing with fat; 'flash sheets' by mixing with collodion and casting on glass plates. These compositions bear a close resemblance to the incendiary bomb and to Thermit welding mixtures, which reach a temperature of over 2000°C.

To minimize risk of accident, the metal powder and the substance supplying the oxygen were packed separately and mixed just prior to use. The powder was fired on open trays with a built-in 'flint' sparker to ignite. Flash powders were also supplied in capsules fired electrically, which gives the ability to synchronize with the camera shutter at 1/10 to 1/25 second.

Flash powders soon gave place to the flash bulbs of amateur and professional photography. Early forms were about the size of a domestic light bulb, containing a crumpled sheet of aluminium foil in place of a filament. The atmosphere in the bulb was oxygen at low pressure. Ignition was achieved by the current from one or two dry cells and total time of the flash was about one-twentieth of a second. Synchronization of flash with camera shutter became possible. Modern flash bulbs are much more compact, contain metal wire rather than sheet. The radiation from these sources is very close to that of a black body radiator and colour temperature may correctly be expressed. Typically it is about 4000K, up to 6000K if 'blued' by using coloured glass for the envelope.

Electrical incandescence and filament lamps

The sources so far described in this section depend upon incandescence brought about by the chemical reaction of combustion. It is also possible to raise the temperature to produce incandescence by passing an electric current through a suitable wire.

The function of the current is purely that of raising the temperature to that required to reach the incandescent state. Such a source is converting electrical energy to heat, and the amount of absorbed energy which re-appears as light is relatively small. Despite inefficiency in converting electrical energy to visible radiation, the filament electric lamp is frequently preferred to more efficient sources on such grounds as simplicity and availability.

An electrical conductor such as metal wire always has some resistance to the flow of an electric current. It is the property of resistance which makes the incandescent filament lamp possible. Electrical pressure or force is measured in volts; when a sufficient voltage or pressure is applied to the conductor, more or less current flows in accordance with Ohm's law. By a correct choice of voltage and resistance any required temperature is attainable. Hence the 'red heat' of the electric fire and the 'white light' of the lamp; likewise the melting of the fuse wire when the current in a circuit becomes greater than is safe.

For use in electric lamps, the filament must have a very high melting-point. The radiation produced by the glowing filament is decided in the main by its temperature and not by the material of construction; it functions as a 'black body', giving a continuous spectrum radiation which can be expressed accurately in colour temperature units. The earliest filaments were of carbon, operating in an evacuated glass envelope because the glowing carbon thread would rapidly oxidize in air. A high temperature is required to give as much visible radiation as possible, but radiation is limited by evaporation of the carbon and its deposition on the inside of the envelope. About 2000K could be achieved.

Eventually tungsten wires suitable for this purpose were made and had the advantage of reaching a higher temperature, and thus giving more visible radiation, than carbon filaments (2400K; tungsten melts at 3655K). Later it was found possible to operate tungsten filaments in an atmosphere of inert gas rather than in a vacuum, and attain a colour temperature nearer 3000K. The most useful gas was argon, with some admixture of nitrogen. During use, the tungsten slowly evaporates and forms the brown stain on the glass bulb which indicates that a lamp is nearing the end of its useful life. As the brown stain builds up on the inside surface of the glass, it progressively absorbs some of the radiation from the filament, not only reducing the amount of visible radiation emitted but also in effect lowering the colour temperature (i.e. the light is depleted in blue light, the shorter wavelengths).

Tungsten-halogen lamps

A very important introduction of recent years has been the tungsten-halogen lamps. In these the evaporation of tungsten on to the glass (or quartz) envelope is prevented. It becomes possible to operate the lamp at a higher filament temperature. The radiation of the tungsten-halogen lamp is still that of the tungsten filament at the operating temperature. Its advantages are a higher operating temperature, with higher colour temperature; greater efficiency with longer life; and smaller size; but no fundamental change in the nature of the tungsten filament radiation occurs, other than the characteristic of the operat-

57

ing temperature which may be up to about 3400 K. The function of the halogen (usually iodine) is basically similar to the role it performs in mercury sodium arcs (see page 62). In this case, the tungsten evaporated from the filament which would normally be deposited on the envelope, reducing its light transmission, is re-deposited on the filament.

Ordinary tungsten filament lamps for domestic purposes – up to say 100 watt – operate at about 2600–2750K. Compact filament lamps as used for projection purposes operate at a little over 3000K, and are shorter-lived as well as more expensive.

Overrun tungsten lamps

Tungsten filament lamps, as 'black body' radiators, emit a great deal of infra-red radiation and proportionately more red and yellow light to blue than appears in daylight. A filament temperature of 4500K upwards would be necessary to approach the spectral com-

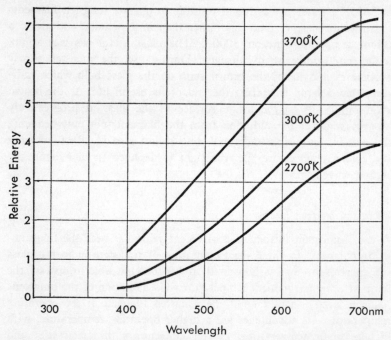

Spectral energy distribution of radiation from gas-filled tungsten filament lamps: normal lamp 2700K; projection lamp 3000K; overrun lamp 3400K (2–6 hour life).

position of daylight and obviously this is not possible with tungsten on account of its melting-point. However, if a tungsten filament is overrun, that is supplied with a higher voltage than is normally intended, a higher filament temperature is obtained and therefore a greater proportion of shorter wavelengths (blue) occurs. The overrun lamp is much more useful photographically for this reason (photographic materials being basically blue-sensitive). Operating at 3435K approximately doubles the radiated energy in the blue region. The other effect of running at high voltage is that the life of the lamp is much reduced – say to some 2 hours of (continuous) runnning. There is less shock to the filament if voltage is run up from normal by variable resistance or variable transformer. The simple construction of the tungsten filament lamp makes it possible to have the advantage of higher colour temperature at relatively little outlay.

The reverse situation to overrunning – low voltage supply – causes loss in blue radiation and a more yellow, even red, light. Mains supply voltage can easily fluctuate by plus or minus 5 per cent. This can raise or lower colour temperature by 75K.

Discharge lamps: radiation from gases

When describing practical light sources relying on incandescence, it was possible to deal in general terms with all types, because the radiation is mainly characteristic of the temperature of the radiator and not on its material of construction.

The description of practical discharge lamps has no comparable simplification, because each gas used in a lamp radiates in its own characteristic fashion. There would be no point in presenting here the radiation characteristics of all known gas discharges; those which are of interest in reprography are based upon the use of mercury, sodium, and xenon. These find practical application in:

General illumination and street lighting: mercury and sodium types
'Safe' lighting in reproduction departments: sodium
Projectors (compact source): mercury, xenon
Photo-copiers: mercury
Printing-down frame: mercury and xenon
Camera copy-board illumination: xenon.

Many different types of discharge lamp have been designed to fit the special requirements of different situations, and it will not be possible to give details. Specialized information is mainly required by

59

the engineer designing equipment rather than the user, and manufacturers of these lamps are of necessity generous in providing details for this purpose.

From the user's point of view, it may be important to know the type of lamp installed in an equipment so as to select the best for the purpose, or to use it to best advantage. Discharge lamps come in many different shapes and sizes: spherical, tubular, helical, etc. The discharge proper – that is, the radiating column of gas – is often contained in an inner envelope of glass or similar material, and this again in an outer envelope of glass. The electrical connections may be through a single cap, or at each end of the tube. Usually some kind of electrical gear is required – transformers, chokes, condensers, etc – to give correct conditions for starting and running. Naturally it is desirable to simplify control and operation of these lamps; but whatever the cost and complications, their wide use today is ample evidence of advantages over other sources.

Principles of discharge lamp

Knowledge of electrical discharge in gases dates back over some 250 years, but no practical form of light source was available until the early years of the twentieth century. Early discharge tubes were laboratory items leading to important discoveries of atomic structure and behaviour, but the low *intensity* of the visible radiation ruled out their use for lighting purposes.

To achieve a radiation of more useful intensity, it was necessary to arrange that the discharge should absorb more electrical energy for conversion to light. This has been achieved by much improved materials and methods of construction, and by using higher operating pressures of gas within the lamp. Hand-in-hand with these conditions goes a higher working temperature; at really high pressures and temperatures, the radiation obtained becomes progressively nearer (but never the same as) the radiation of an incandescent solid. More precisely, the line or band discontinuous spectrum of the simple gas discharge acquires a 'background' of continuous radiation, on which the basic line spectrum is superimposed as 'peaks' of higher intensity.

Thus it is common to speak of 'low pressure' and 'high pressure' discharge lamps. To give these terms some significance, low pressure indicates some small fraction of atmospheric, while high pressure is applied to about and above atmospheric. For instance, the Cooper-Hewitt mercury vapour lamp which appeared in the early 1900's, operated at a pressure of only 1 to 2 millimetres (atmospheric pressure is 760mm.). This was the first successful discharge lamp for lighting purposes, and found use in photographic studios. Luminous efficiency was low by modern standards, but in the days of black and white photography and predominantly blue-sensitive emulsions, it was welcomed as making possible reduced exposure times.

The simple discharge tubes of the laboratory relied on platinum wires sealed into the glass walls to convey electricity to the 'electrodes' inside the tube. Much progress has been due to better methods of making this glass-to-metal seal, more suitable glasses for the purpose, and a variety of developments of the electrode itself. Compared with tungsten filament lamps, discharge lamps are more expensive to purchase and install and are more easily damaged. They also require a closer adherence to the stipulated running conditions if they are to give the intended performance; this is particularly the case where the non-visible radiation is concerned. For this reason it is becoming increasingly common to include voltage stabilizers in the equipment.

One of the features of using some discharge lamps is a 'running-up' period between switching-on and reaching full radiation performance. With mercury and sodium lamps, this stems in part from the fact that in the cold lamp mercury is liquid and sodium is solid. Gas discharge can take place only when they are at least partly vaporized, which is effected by an auxiliary circuit or by the inclusion of a small amount of an inert gas to initiate the discharge. In this case the lamp when switched on starts as a discharge in gas at a low pressure, but as the metal is vaporized and reaches the working pressure the amount of radiation due to the initiating gas becomes insignificant.

A further inconvenience with discharge lamps is that in addition to warming-up, they require a cooling-down period before re-starting. For reprographic purposes, the chief effect is to make it necessary to control exposure time with shutters rather than by switching lamps on and off. Despite the disadvantages mentioned, their use is justified by high luminous efficiency (proportion of absorbed energy converted to useful radiation), long life under correct operating condition, adaptability to a wide variety of applications, and the ability to choose from a range of spectral energy distributions.

Mixed vapour discharge lamps

Current developments in discharge lamps are concentrated on the use of mixed vapours, mainly with the objective of a more nearly

61

white light. From the point of view for instance of street lighting, mercury vapour alone or sodium vapour alone is unpleasant on account of poor colour appearance of various objects: the blue-green discharge of mercury is deficient in the red and yellow regions, while the amber light from sodium lacks blue radiation. Attempts are being made to combine the emission of two or more substances in the one tube, to give a more balanced emission through the spectrum. Note that it is not necessary to have a continuous spectrum like that of daylight to give the visual effect of white light. So long as the retina of the eye is activated equally at the appropriate wavelengths, the brain receives the stimuli which it interprets as 'white'. In a reprographic situation, it is not sufficient that light *appears* correct to the eye, it must have the correct spectral energy distribution for the purpose in hand.

The attempt to modify the mercury emission by incorporating sodium in the same lamp is an example of the type of problem encountered in this field; the ordinary sodium lamp operates at low pressure and a low temperature – about 530K. If sodium is to contribute usefully to the mercury emission, it would have to operate at higher pressure and temperature – about 2500K. At this temperature, sodium vapour is highly reactive and chemically attacks any of the glasses which might be used for containing the gases. Therefore further progress has had to wait for new constructional materials to be developed. One such new material, announced within the last ten years, is named Lucalox. It is not affected by the vapour of sodium and other metals at temperatures of over 1700K. High-pressure sodium lamps using Lucalox were announced in 1960, giving much improved colour rendering.

Another development of great significance comes from the addition of other metals to mercury vapour, but in the form of their iodides. This makes possible a greater effect on the radiation while preventing chemical attack on the glass envelope of the lamp. As with the tungsten filament lamp, part of the benefit comes from the temperature gradients within the envelope; the lower temperature near the glass walls protects the glass from attack, but in the arc itself the metal iodide is 'dissociated' and the metal atoms can be excited alongside the mercury atoms. In fact it becomes possible to depress the mercury radiation to a subsidiary role. This technique has made it possible to add sodium, cadmium, lithium and zinc to mercury with great improvement in the visual output. There are indications that white light may be obtained from the addition of thorium to mercury

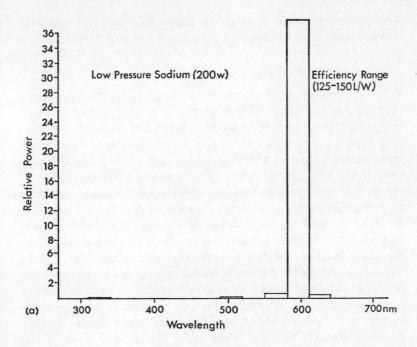

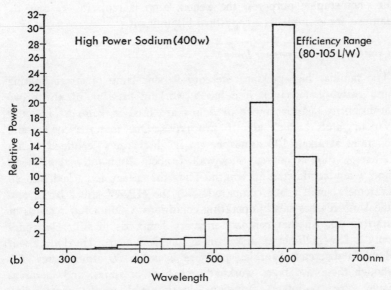

Change in emission of Lucalox sodium vapour discharge lamps: *a*, low pressure, 200 watt; *b*, high pressure, 400 watt. From "Recent developments in Discharge Lamps" in *Light and Lighting*, June 1966, p. 167.

– an interesting point to reach in comparison with the value of thorium oxide as an incandescent light source (see pages 54 and 68).

Xenon discharge lamp

While these latest developments of discharge lamps may eventually have far-reaching effects in reprography, they have not yet 'arrived' in an acceptable practical form. In contrast, the high-pressure mercury vapour lamp has developed to a very reliable source for photo-reproduction purposes; effective, robust, and of long life. Its only rival is the xenon lamp which has the obvious advantage that its spectrum is continuous through the visible region. Although this gives the xenon emission a close approximation to daylight, its energy in the photochemical region is relatively low. In a photoprocess employing materials whose sensitivity is mainly in the ultra-violet region, the xenon lamp is providing a useful amount of actinic light but its emission in the visible regions is wasted. Its advantage over mercury is that of switching on and off instantly – xenon is a gas at ordinary temperatures so the discharge can start as soon as the electricity supply is connected. The control gear for xenon lamps is more expensive than for mercury. When white light is required for visual or photographic purposes, the xenon lamp is replacing carbon arc lamps – for example, as a copy-board illuminant.

Compact mercury vapour lamps

The familiar high-pressure mercury vapour lamp of street lighting and early photocopying use has a working pressure of about two atmospheres. Much higher pressures are used in *compact* mercury vapour lamps, which are of two types. One is water-cooled and operates at about 100 atmospheres; its discharge is confined within a narrow tube of quartz, thick-walled, about 2mm internal diameter and a few centimetres in length. Thus the energy is packed into an extremely small space compared with the HPMV street lamp, and the lamp reaches steady operating conditions within a few seconds of starting up. Another compact mercury lamp has massive electrodes only a few millimetres apart, in a spherical quartz envelope about 4cm. in diameter. Working pressure is up to 20 atmospheres. Although these two types work at such high pressures, and therefore have some continuous radiation superimposed on the normal line spectrum, their radiation is not suitable for photography if good colour rendering is required.

The small sizes of these compact mercury vapour lamps makes it possible to use them in optical systems where the radiated light requires to be collected and directed. Typical are focusing systems, for localized lighting or image projection. At least one similar source using xenon is available for an 8mm. film projector and has the obvious advantage of close approximation to daylight; but in reprographic applications when a compact xenon source is required it is more usual to form this as a helical tube. Apart from the physical shape of the tube, the radiation is identical with the typical xenon lamp which has a straight tube about 14in. long.

Ignoring the use of gas discharge lamps for lighting as such, the application of these lamps to reprography is overwhelmingly in the form of tubular mercury vapour lamps used in modern rotary photocopiers; and to a lesser extent, in mercury vapour and xenon lamps for printing-down purposes in flat frames. Discharge through mixed metal gases, mentioned on page 62 in connection with visible emission, may be equally valuable in 'tailoring' the lamp to the reprographic application.

Categories of light source

The sources so far reviewed have fallen into one of the three categories: effects in cold solid materials, in hot solid materials, and in gases. It should by now be clear that there is no hard and fast dividing line between one such group and another. Each radiant depends upon molecular excitation as its means of converting some form of energy to radiation; and the nature of the radiation is decided by the molecular state of the radiating substance. But from a practical point of view, the continuous spectrum radiation of incandescent solids can be separated from the discontinuous radiation of the majority of gases.

There are a further two groups of sources of especial importance in reprography, which involve the combination of two of the three effects simultaneously. These are carbon arc lamps and fluorescent tube lamps.

Carbon arc lamps

In the early days of the filament lamp, an 'electric candle' was devised (1876) in which two parallel rods of carbon were separated by a wall of porcelain. It was possible to strike an arc between the tips of the two rods, and in use the porcelain vaporized as the carbon rods shortened by 'burning away'. From this primitive beginning, developments were in the direction of general lighting, until the filament lamp established itself in this field. Arc lamps then found their major uses in photographic applications and a considerable technology was built up. There is a movement away from arc lamps today on account of certain inconveniences, but they still hold a special place in photomechanical work, and where a very compact high-intensity source is required in an optical system; examples are in the cinema projector and the searchlight.

It is usual to speak of low-intensity and high-intensity arcs, open and enclosed arcs, and flame arcs. Additionally, there are differences due to operation on DC (direct current) or AC (alternating current) electrical supplies. The voltages used are low relative to other sources, some 40 up to 80 volts being common. Current consumption can be high, i.e. energy absorption can be large and light emission correspondingly high.

An arc lamp is started, or struck, by connecting the two carbon rods to the electrical supply and bringing them into contact. The resistance to electrical current where the rods are in contact causes some heating. Then the rods are separated, at first by only a short distance. The current continues to flow but now has to jump across the gap between the rods, hence 'arc'. Further heating of the rods occurs, with some vaporization, and the distance between the tips can be increased until a steady working condition is obtained.

On a DC supply, one of the carbon rods is referred to as the 'positive' and a crater forms in the tip. This crater is the major source of radiation in a high-intensity arc as used for projection purposes. The carbon rod would be some 10–15mm. in diameter, which therefore defines the size of the source. As the carbon rods burn away, the correct distance must be maintained either by hand or some semi-mechanical arrangement.

In the steady condition, the walls of the crater on the positive electrode become heated to some 4000K. They therefore radiate with a continuous spectrum as typical of a black body at that temperature. Additionally, the crater is filled with a mixture of carbon particles and of hot gases. The temperature of the gases in the crater is 5000K or above; thus there is superimposed upon the continuous radiation of 4000K colour temperature from the walls, the continuous radiation from carbon particles at 5000K, and the discontinuous radiation from molecules of gases under the excitation of the electrical field between the electrodes.

66

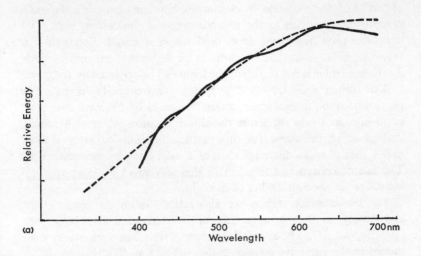

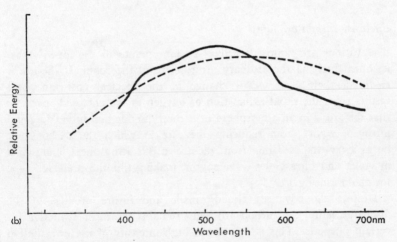

a, Radiation from low-intensity carbon arc lamp (solid line) in comparison with black body radiation at 400K (broken line). *b*, Radiation from high-intensity carbon arc (solid line) compared with radiation from black body at 5400K (broken line). Jones and Crabtree, JSMPE **10**, 146 (1927).

High-intensity and flame arcs

With the high-intensity arc, it is the small size of the crater relative to the energy emitted which is desirable in conjunction with the lenses or mirrors of the apparatus. Where this is not necessary, the electrodes can be separated by a greater distance; this increases the length of 'arc' and the radiation owes more to the emission from the gases carrying the arc than to the incandescence of the carbon rods. This condition gives the flame arcs, used where a broad illumination is required rather than a concentration of light into an optical path. As the hot crater is not required, the lamp can be operated on A.C.

The carbon rods for these purposes are frequently 'cored'. The purpose may be to encourage crater formation by having a soft core; or to introduce into the flame metallic substances which will modify the colour of the flame (to this extent, the emission is then more nearly that of a gas discharge than of a solid body by incandescence). The use of cerium and thorium in this way gives a closer approximation to the spectral quality of daylight.

The low-intensity carbon arc also relies less on the crater effect, and mostly on the incandescence of its electrodes, especially when operating on AC. The spectral diagram of this emission is very close indeed to the curve for a black body radiation at 4000K.

Enclosed carbon arc lamp

The various arc lamps described so far operate in the 'open' – that is, burn freely in the ordinary atmosphere of the room. If the arc is *enclosed*, there is a useful change in the spectral emission. This results from the rapid exhaustion of oxygen in the enclosed space, so that the arc is in an atmosphere of carbon dioxide and monoxide, and nitrogen. Apart from reducing the rate at which the carbons are burnt away, the radiation from the arc in this atmosphere is enriched in violet and ultra-violet wavelengths, making this the preferred type for photoprinting use.

Despite the advances already made and future promise of the discharge lamp, the carbon arc lamp remains the most attractive for certain purposes. This is due to the high intensity of useful radiation (visible and ultra-violet) which can be obtained from a relatively small source. As will be discussed later in the context of precision photoprinting arc lamps are commonly described as 'point sources'. The disadvantages of arc lamps are that they require a mechanism for striking the arc and for feeding the carbons as they burn away; that carbons must be replaced fairly frequently; and that they produce dust (particularly undesirable in modern establishments seeking to maintain conditions of high cleanliness). Enclosed arcs also suffer from the deposits formed on the inside of the glass. These are highly absorbent of ultra-violet radiation, and must be removed frequently.

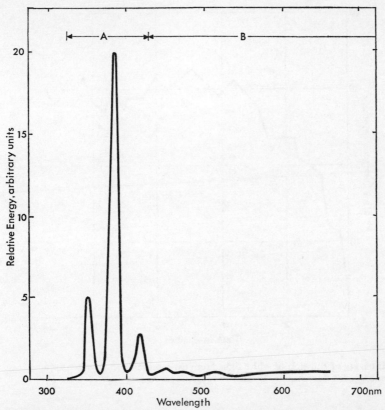

Distribution of energy in the emission from an enclosed flame arc. The strong peaks in the region A result from enclosure of the arc. Cooper and Hawkins, "Spectral Characteristics of Light Sources", JSDC **65**, 586 (1949).

An advantage is that arc lamps can, within limits, be switched on and off with fairly immediate response, but unfortunately not if the exposure required is of short duration – say, less than five minutes (on account of a settling-down period). Even when an arc lamp is running 'steadily', there will be fluctuations which prevent their use when repeated exposures of a critical nature are involved. In printing-down using large frames, times are frequently 15–20 minutes and these fluctuations are then of less significance. Most large shops relying on arcs avoid this particular problem by using an integrating meter, by means of which exposure is controlled not by time but by the summation of dosage as measured by a photo-cell and recorded on an arbitrary scale.

69

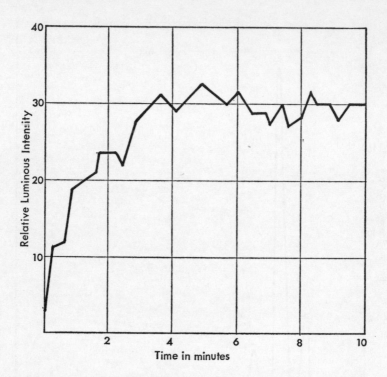

Intensity of emission of white light from carbon arc lamp on start-up and subsequent fluctuations. Wallis, *Process Engravers Monthly* **61**, 41 (1954).

Fluorescent-tube lamps

The second group of light sources of mixed type is the fluorescent tubes, now widely used for domestic and industrial lighting, and to some extent for street lighting. These tubes are low-pressure mercury vapour discharge lamps; but the discharge takes place in a glass tube which has a coating of fluorescent material on the inside. Certain patterns are especially interesting for reprographic purposes.

It will be remembered that at low pressures, the emission of a mercury vapour discharge is a discontinuous, or *line* spectrum, predominantly only a few strong lines in the visible regions. There is also emission at invisible wavelengths, which is wasted in the ordinary way and may never get beyond the glass walls of the tube owing to the opacity of glass at these wavelengths. Some of these radiations can be absorbed by the fluorescent coating, and re-emitted at wavelengths in the visible region to which the glass is more transparent.

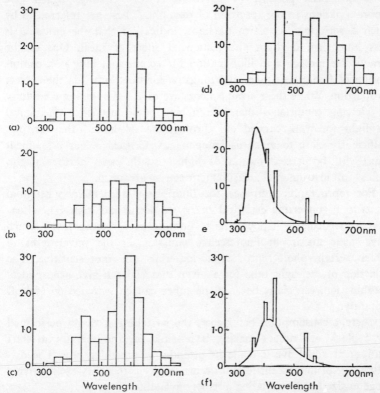

Spectra of low-pressure mercury vapour discharge in various fluorescent tubes: *a*, daylight. *b*, warm sunlight. *c*, warm white. *d*, experimental colour matching. *e*, super actinic, Osram Type 70. *f*, photo-blue ("actinic"), Osram Type 71. From Ruff, Trans. Ill. Eng. Soc. (London) **29** (4), 129 (1964) and Osram G.m.b.H., Berlin.

(See the list of fluorescent chemicals on page 38). The total effect of this combination is to:

1. Reduce the dominance of the major mercury lines
2. Convert non-useful radiation to useful
3. Give the line spectrum a continuous character.

The result is therefore quite different from that obtained by operating mercury lamps at higher pressure: this gives some continuum (or white light) but the characteristic lines of the mercury spectrum are not diminished in effect, nor is there any less wasted radiation. Also, when other metals are added to the mercury vapour, although there is significant improvement the spectrum remains a line or band spectrum.

71

It has been possible to select fluorescent coatings which give approximations to the spectrum of daylight. These are referred to by such descriptions as warm daylight, indicating that the emission is less blue than daylight and somewhat short of ideal. One of the practical approaches to illumination for colour-matching is a combination of fluorescent tubes and tungsten filament lamps in the correct proportion. While these arrangements are of value in giving continuity of viewing conditions – that is, reducing our dependence on natural daylight for work carried out through the 24 hours – they are not sufficiently close to make us independent. Critical colour assessment must still be carried out in daylight, usually noon north skylight, unless full instrumental measurement can be arranged.

For reprographic purposes, the fluorescent lamp offers substantial benefits in converting electrical energy to useful radiation. The situation is particularly happy because fluorescent substances exist which have their maximum fluorescence emission at the wavelengths to which certain photo-reproduction materials are most sensitive. The selection of the right tube for a given material can give a near-ideal combination, especially because the tubes can be switched on and off instantly.

Current consumption being low, the switching is easily performed by a clockwork timer – making it possible for the operator to start equipment and leave it while he attends to something else. The disadvantage of fluorescent tubes in reprography is that they are relatively large in size for the radiation intensity available.

For this reason it requires a large number of tubes at a short working distance to give the level, or intensity, of radiation necessary to affect reprographic materials in a reasonable time. This is the reverse of the case with high-pressure, compact sources, and makes the one serious limitation to the use of fluorescent tubes in critical reproduction systems. Their attractions remain high for the reasons given above, and they deserve a greater degree of consideration than is normal at the moment. Exposure times can be very short in comparison to the low energy consumption.

Selection of light sources

In the home, office and factory the usual choice lies between tungsten filament lamps or fluorescent tubes. The first issue is between initial cost and running cost; tungsten is the cheaper to install, because it merely requires connection through suitable fittings, wiring and switches to the mains supply. Fluorescent is more expensive initially owing to the cost of the lamp, the necessity of putting the tube into a fitting of some elaboration, and the need for control gear in the circuit (usually in the fitting itself). The fittings for fluorescent tubes even of small size – say 20in. length – are considerably more bulky than the majority of fittings for tungsten filament lamps; therefore space limitations might decide against fluorescent tubes in favour of tungsten lamps. But if these factors are not of major importance, the logical choice would be for the system which is more *efficient*, that is, better in converting energy consumed into useful radiant energy.

Invariably some of the consumed energy reappears as heat. From the illumination point of view, the heat generated may serve no useful purpose; it may even require that special cooling arrangements be made so as to prevent the apparatus becoming overheated. Electrical insulation is easily damaged by high temperature – for example, the rapid embrittlement of rubber in the leads to an electric iron.

The absorbed energy which does not re-appear as heat is visible as light, or can be detected as invisible radiation. For some purposes the invisible radiation may be more important than the visible light. The relative value of alternative sources cannot be assessed unless we have some means of measuring the amount and type of the radiation which cannot be detected by eye. The eye itself is a very sensitive instrument with respect to quality (colour) and quantity (or brightness) of light. Remember, however, that what we 'perceive' is the brain's interpretation of the signals sent to it as a result of chemical changes taking place in the retina. The ability of the brain to interpret is greater than commonly supposed. For example it is a matter of everyday experience that colours cannot be judged (matched) reliably under artificial lighting; 'daylight' fluorescent tubes have been designed and the illumination they provide is very much closer to daylight than tungsten filament lighting. Yet the brain's capacity for adjustment is so great that some people *prefer* to view colours under tungsten light rather than under daylight-matching fluorescent light. This is because the individual's brain has learnt the degree of compensation to make under the less perfect conditions. Of course a person whose experience or mental training has been gained under some other condition might well have another preference.

The point is that even after all the considerations of initial cost, efficiency and running costs, physical limitations or aesthetics of design,

Conversion efficiency of lamps

	A Heat %	B U.V. Radiation %	C Visible Radiation %	D Infra-red Radiation %	E A + D %
1. 100w Tungsten	19	0.25	5.75	75	94
2. 80w low-pressure mercury	24	56	5	15	39
3. 2kw Xenon	25	2	15	58	83
4. 400w high-pressure mercury	28	9	11	52	80
5. 400w fluor. high-pressure mercury	29.5	1.5	14	55	84.5
6. 200w low-pressure sodium	41.5	0.5	22	36	77.5
7. fluorescent tube	45.6	2.4	22	30	75.6

there may still be an overriding *preference* for a system which is unsuitable by some other standard.

Another example of the adaptability of the visual process is the fact that if a white surface – say a sheet of paper – is viewed in relatively yellow candle-light, it is seen as *white*, not yellow. This compensation takes place quite unconsciously.

Efficiency in converting energy to radiation

The table opposite shows data taken from *Discharge lamps: Colour, efficiency and application* by H. R. Ruff (Transactions of the Illumination Engineers' Society, Vol. 29, p. 129, 1964). The tables give as a percentage of the energy absorbed, the way in which it re-appears. For this purpose, the breakdown is into actual heat, visible and invisible radiations, the latter including ultra-violet and infra-red. The information is arranged here so as to present the various illuminants in ascending order of heat production (Column A); so that on this criterion *alone*, the most efficient source tops the list, the least efficient comes at the end.

All the illuminants listed in the table produce a considerable proportion of heat and a good deal of infra-red radiation. The differences as to ultra-violet radiation and visible radiation are, however, very much greater than the heat and infra-red differences. If our interest was the generation of heat as such or infra-red radiation for heating effect, all of these sources could be considered eligible on grounds of energy conversion efficiency; the least efficient being No. 2 and the most efficient, No. 1 (Column E). The differences in visible radiation are greater than the heat plus infra-red, but the differences in ultra-violet radiation are greater still: a range of 0.25 per cent from No. 1 to 56 per cent from No. 2.

Fluorescent-tube efficiency

The fluorescent tube is particularly interesting as an energy converter. For example, a tube absorbing 80 watts initially as an electric discharge in mercury vapour, emits or converts 48 watts into ultra-violet radiation and only 2 watts to visible light; the remaining 30 watts is lost as heat, either at the cathodes of the tube or along its length. Of the 48 watts appearing as ultra-violet radiation, and absorbed by the fluorescent coating of the tube, 12 watts is re-emitted as visible light; 23 of the 48 watts re-appears as infra-red radiation and 13

75

watts as actual heat carried away by conduction or convection currents. The 80 watts absorbed therefore reappears as:

2 watts direct visible light production
12 watts visible light by fluorescence
23 watts as infra-red radiation
43 watts as conducted or converted heat

Questions of terminology

Up to this point we have made rather casual references to ultra-violet and infra-red. These terms are convenient in common use but greater precision is frequently necessary. Earlier the *wavelength scale* was used to define the various regions of the visible spectrum as an alternative to colour. It is important to realize that the names applied to colours or to the spectral regions either side of the visible spectrum are merely terms of convenience; it is in fact quite erroneous to speak of *the* ultra-violet as if it were a clearly recognized and finite thing – the phrase merely indicates in a loose way, the region found beyond the blue and violet and by definition, not visible to the human eye.

The majority of reprographic processes rely upon chemical changes which are brought about by radiations of a restricted wavelength region. But whereas each human eye is sensitive to radiation down to some definite point on the wavelength scale and not beyond that point, there are processes which have sensitivity both within the visible region *and* beyond the visible into the ultra-violet. Other processes or chemical actions are completely 'blind' to the visible region and show sensitivity in areas to which the human eye has no response. Sometimes the slightly more descriptive phrases 'near ultra violet' and 'far ultra violet' are used, but again these are too imprecise for systematic study of photoprocesses. In the succeeding sections we shall be considering radiations in terms of their chemical effects and it will be necessary to give a picture of precisely where these effects arise. For this purpose a graph or curve is again more informative.

Another term in common usage is 'actinic light' by which is meant radiation to which a reprographic material responds. Conversely, 'non-actinic' implies radiation to which the material shows no response. This also is necessarily a loose and vague term because the various reprographic processes differ in their responses. Thus light which is actinic to a silver-sensitized film may have no effect at all on a diazo-sensitized product. This makes it possible to handle the diazo product under lighting conditions which might be unsafe for

76

the silver material. Hence the terms actinic and non-actinic are only meaningful with respect to one specific light-sensitive system.

Value of spectral diagrams

Thus the everyday language used in reprography contains terms and phrases which are not exact in their meaning. The alternative, of always specifying precisely the conditions, would make communication difficult. Here again the spectral diagram overcomes the problem by presenting the facts in visual form. Spectral diagrams might be called the 'language' of reprography.

Most of the spectral diagrams shown in this book have been re-drawn to the same size and with a common wavelength scale. This makes it easier to transfer mentally the facts of one diagram on to another – as is necessary in, say, comparing two light sources for possible use in exposing a given sensitive material. In practice there are other factors than these to consider but as the basic reprographic situation is the relationship between radiation and radiation-sensitive material, it is well to become familiar with the various sensitive systems at this stage.

In using spectral diagrams for this purpose, it is the shape of the curve which matters. All are based on a horizontal scale of wavelength, and a vertical scale of *percentage*. Of course the vertical scale would equally, or for some purposes preferably, be on a quantitative basis. The *actual quantity* of energy radiated at each wavelength could be shown, or some measure of the *actual* sensitivity of the material used; but this is not necessary for an appreciation of the reasons why a given reprographic situation arises.

The percentage scale takes the point of maximum effect as 100 and expresses all other points on the curve as a proportionately lower value, e.g. 50 for half as much effect, 25 for a quarter and so on. This has already been referred to in describing the spectral sensitivity of the eye; the same principle is involved in defining the sensitivity of a reprographic material.

CHAPTER 4

PRINTING METHODS

Before moving on to consider photoreproduction processes, it would be as well to look at the relationship between printing proper and photoprinting; not only because the IRT syllabus requires some knowledge of printing methods, but because there is a common dependence on photochemistry. Photoreproduction has a *direct* involvement in photoprinting, and an *indirect* involvement in the printers' activity. To rephrase this, photoprinting is an end in itself; but to the printer, photochemistry is a means to an end. In considering photosensitive systems we need not distinguish between the two fields of application, for first emphasis is on the system and not how or where it is used.

This section then presents information required by the syllabus with respect to the printing processses, and shows the place of photoreproduction mechanisms in those processes. It will be seen that the development of photoreproduction systems has inextricably been part of recent development in the older crafts of printing; while the indications are that the future may make it difficult to see where the one ends and the other begins.

Mechanical printing and photoprinting

By and large, the printing processes as commonly understood are mechanical systems for producing a large number of copies. In printing, the main concern must be with the mechanical or machine aspects of the operation. It is in the nature of machines that they need skilled operation to avoid damage and give their best. They take time to set up and the effort of doing so is justified by the number of copies required. This number varies from the daily newspaper with an edition of millions to the personal cheque book requiring only hundreds of impressions. In each situation there is a variety of processes from which to choose, but one tends to use the equipment or method to hand rather than one which has to be bought in specially, borrowed, or 'hired' on another man's premises.

Photoprinting achieves a somewhat similar primary objective – the making of copies from an original – but the methods of photoprinting are chemical rather than mechanical. The skill in a photochemical process is more noticeably in manufacture of photosensitive materials. The cost of single copies obtained by photoprinting tends to be high compared with those by conventional printing, but the equipment necessary is usually simpler and less expensive. The tendency is to use photoprinting for smaller runs of copies which can be produced 'on the spot' whereas printing work usually has to be sent out to where the machine and operating skill is available. At the same time, there are no obligatory frontiers or lines of demarcation between mechanical printing and photoprinting; one can cross from one to the other according to the requirements.

Mechanical printing methods have undergone some 500 years of development in the West, though known for considerably longer in China. Photo-printing has had only 125 years in which to develop its methods from rudimentary beginnings to sophisticated techniques. Certain areas of mechanical printing have benefited from the techniques of photoprinting – noticeably in the provision of illustrations but now to an increasing extent in the preparation of text. Conversely, photoprinting has moved from situations where simple equipment was sufficient, to highly-mechanized apparatus which has made use of the printer's ingenuity and experience. Thus a mutual intermingling of ideas and technologies has come about and much common ground has been established.

The photoprinting community cannot avoid the expansion of its interests into areas which had been the prerogative of the traditionally closed community of the printer; as well as into totally new situations brought about by such concepts as the punched card, miniaturization and computer print-out. Rather more belatedly perhaps the printer has recognized that his requirements now range more widely than in previous generations; for him the term 'graphic arts' has had to extend its concern into techniques and materials somewhat alien to the centuries of craftsmanship behind him.

Some have seen in these evolutionary phases a rivalry between the established and the newer arts; but the enlightened view must surely be that each has to learn from the other. Reprography is concerned with processes rather than mechanics; its aim is to arrive at a better understanding of basic process mechanisms, so that practical results may more nearly be the best rather than the one which 'gets by'.

As might be expected, the printing industry is more advanced in such matters as training, specialised education and technological research. In undertaking a broadly-based approach to the elementary scientific facts one can but recognize that it is the older industry which gave some processes the reason for their existence. While, therefore, it is unnecessary to repeat here what is adequately covered in excellent texts prepared for printing technologists, it is important that reprographers have some background information to the processes of printing.

Printing ink – properties and function

All mechanical printing processes involve printer's ink or the equivalent, but methods differ in the means adopted for transfer of ink in a selected form or design to the material which is to be printed. Printing ink also plays a part in some reprographic methods which tend to rely on chemical reactions rather than physical transfer.

Printing processes can be very demanding in the properties required in the ink. For present purposes it may be thought of as a liquid vehicle which carries colour or pigmented material. The vehicle is in the nature of a varnish, sufficiently viscous to stay in position on the printing press and to pass to another position only when required to do so. It conveys colour or pigment in the intended arrangement or design to the final article, where it then has to dry so that it can be handled and to anchor the pigment in its final position. It is not unlike oil paint in some respects, though usually quicker-drying. Printing on paper aids drying through absorption of the vehicle into the porous paper, followed by evaporation and/or hardening of the varnish-like constituents through atmospheric oxidatino (chemical combination with the oxygen of the atmosphere). Printing on metal or other non-porous surfaces relies more completely on rapid evaporation/oxidation and may be assisted by heating the printed articles to quite high temperatures. High-speed printing on paper may also require heat-assisted drying. 'Thermofast' ink relies on increase of temperature to initiate a chemical reaction, whereby a thermo-setting resin is produced in the ink layer without dependence on evaporation or oxidation effects. (Sun Chemical Company, of U.S.A.).

Printing methods

The essential differences between the various printing processes are in the means whereby the ink pattern is created and then transferred to the final surface:

Letterpress or relief printing: The oldest printing method relies on the required design being a *raised surface*. The high parts are inked and the ink transferred under pressure to the final surface. The humble rubber stamp comes into this category. It is the traditional technique of printing, and stems from the earliest hand-cut wood blocks and subsequent hand-cut single letters or movable type. The usual term is *letterpress* but the method is not restricted to letters or type and it is best to think of using a design which is *in relief* for transferring the ink. Being in relief, the printing parts are raised above the non-printing background.

Planographic printing: A more recent method relies on a flat surface, and the design must then be created by compelling some areas to accept ink for transfer while the non-printing background refuses the ink. As the design is *on the same surface* as the non-printing areas, the generic description for the method is *planographic* though in more familiar use, *lithographic*. Initially, lithography involved the design being drawn by an artist directly on to a prepared stone surface.

Gravure printing: Also more recent than relief printing is the *intaglio* method in which the design for transferring ink is *recessed into* the surface or non-printing parts. This has evolved from the artist's technique of engraving lines in metal and filling them with the ink, while cleaning ink from the rest of the printing surface; hence, *gravure* printing.

Screen printing: Finally there is the method of applying ink to the object by forcing it *through* openings cut in a sheet or stencil. Stencilling in an elementary form is in wide use, in crude designs, for lettering on crates and other packages. For more intricate designs the stencil is supported on a fine mesh of silk, nylon thread, or metal wire; hence *screen printing* and the reference to silk screens.

Photomechanical work

In every one of the printing methods outlined, the technique known today has evolved from original work of an artist/designer in creating a printing block or 'plate'. There is a close resemblance therefore to the work of reprography when the task is to reproduce in quantity, the original work of a draughtsman. Photoreproduction processes have enabled printing techniques to be less dependent on the artist to create the actual printing surface which will be used to control the ink transfer. The field of photography applied to the production of printing surfaces is referred to as 'photo-mechanical' work and is carried out in the 'process shop'. This is where the interests of printing and reprography find much in common.

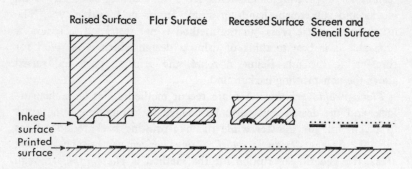

Comparison of the four types of printing. The word *forme*, originally applied to movable type locked into a frame or "chase" ready for the press, is in some countries applied to the printing surface irrespective of the printing method for which it has been prepared.

Inking methods

Printing ink has to be worked up into a suitable consistency by a system of rollers on the printing machine until a uniform film of correct consistency is available on the last rollers. Then, in letterpress printing, the ink rollers pass over the design and transfer ink only to the raised parts which are required to print. In planographic work, the ink on the rollers contacts the whole of the plate surface but is prevented from transferring to the non-printing areas because they are maintained in a wet condition. With intaglio plates, ink is applied to the whole plate surface, filling the recessed parts which form the design to be reproduced. The ink on the non-printing parts is then removed by scraping, only the recesses remaining filled. In screen printing, ink is applied to the whole of the back of the screen by blade but can reach the paper only where the stencil carried by the screen has openings for this purpose. Thus printing, irrespective of the method used, always amounts to a controlled application of ink to the final material.

The printing craftsman has usually become thoroughly competent and expert in one printing method, or possibly some one aspect of one method. The reprographer is not concerned in an attempt to usurp the function of specialist craftsmen, but with a sufficient understanding of the crafts involved for him to make his contribution to the printing exercise when called upon to do so.

Place of design in printing

Every typeface (design of alphabet for printing purposes) has to be designed and drawn by a craftsman. Moreover, the first printing methods used wood blocks on which the written forms of the manuscript were carved by hand. Movable type in the sense of separate characters began the breakaway from total dependence upon the calligrapher and woodcarver, but all typography still depends on a designer-draughtsman who has drawn out each separate character used.

The planographic method also originally required an artist to make his design on a lithographic stone, from which numerous copies could be prepared. Likewise, the gravure technique evolved from the engraving of designs on metal (steel, copper) and the chemical method used for etchings; screen printing first relied on hand-cut stencils. Obviously the supply of artists who can draw on stone, engrave metal, etc., is too limited for the amount of printed material which the civilized world demands; photo-mechanical methods are able to bridge the gap between the work of the individual artist/designer, and the printer. Hence the use of terms such as photolithography, photo-engraving, to distinguish the newer techniques.

Direct and indirect printing methods

With a printing surface and an inking system available, the next step is to achieve the transfer, for which various methods are used. In the first place, the inked surface can act *directly* on the final material; this requires that the design reads 'backwards' in order that the final impression is correct-reading. The inked surface may also be caused to act upon some intermediate surface, the ink then undergoing a second transfer to the final surface. This gives an indirect transfer method, usually spoken of as offset, and the design must be correct-reading on the printing plate in order to finish correct-reading on the final copy.

The offset principle has come into very wide use in connection with planographic printing, so much so that to some people the term offset has no meaning outside lithography. It should be borne in mind that direct and indirect ink transfer can theoretically be employed with any of the printing processes, and that for complete clarity one needs the qualification of 'direct' or 'indirect' in every case.

The indirect or offset principle involves a first transfer of the ink pattern to a surface which is in the nature of a temporary support: for it merely serves to carry ink from the printing pattern on to the final article. The material which has been found best for this purpose is some form of smooth rubber, re-inforced in various ways for the necessary mechanical strength. It is referred to as a 'blanket' but carries little resemblance to the domestic article. Its significance is that the rubber reduces wear on the plate image and permits a resilient pressure contact with the final surface which is to receive the ink. Thus the offset principle has made possible printing on to metal as in widespread use today for packaging purposes or domestic and industrial articles. The blanket allows pressures of the same order (70–75 lb. per square inch) in lithography as in letterpress, but without subjecting the lithographic printing image to these conditions.

Forgetting temporarily the various routes by which printing machinery has evolved, it can be seen that advantages lie with the offset principle applied to both letterpress and lithographic techniques; moreover that the very distinction between these two methods is historical rather than fundamental in significance. An offset printing press which could accept printing-plates from the simplest 'surface' plates through to relief plates of varying height of image offers great flexibility as to type of work which can be undertaken. Such presses are in fact now designed and a few have been installed in Britain in recent years. The photoreproduction processes have been in part instrumental in bringing about this significant development, by making it possible to produce so many types of printing plate by photographic methods; these can become the common denominator of the two printing processes in this type of machine. It is of course necessary to have adjustments which allow plates of varying overall thickness to be accommodated.

Printing pressure and ink transfer

In direct printing from a raised surface, the inked type is applied to the paper under a sufficient *pressure* to transfer the ink. Obviously, requirements vary according to the particular paper and ink used. One of the characteristics of letterpress is that the quality of the work depends upon the pressure applied. It must be sufficient to give filled letters without significant damage to the paper by embossing. Excess pressure also causes the ink to spread beyond the outline of the design, while too little pressure leaves the ink 'thin' in places where it prefers to remain on the type when the pressure is released. Purists find the quality of letterpress printing by skilled craftsmen unsurpassed by any other method and they may be right; but offset lithography, expertly carried out from photo-typeset material has established as high a quality level in general printing work.

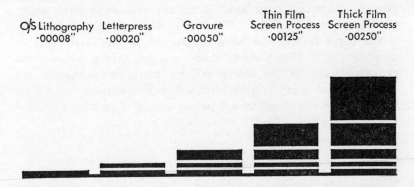

| O/S Lithography ·00008" | Letterpress ·00020" | Gravure ·00050" | Thin Film Screen Process ·00125" | Thick Film Screen Process ·00250" |

Ink film thickness according to printing process. From *Inklings*, No. 64, Coates Bros. Inks, Ltd.

Transfer of ink in planographic printing can also be direct to paper but generally lower pressure is necessary. The raised printing surface of letterpress with withstand more wear and tear than the printing 'image' of a lithographic plate. Intaglio printing involves higher pressures, because the paper must be forced to enter the inked recesses in order to pick-up the ink and achieve the necessary transfer. Screen printing involves the least problems with pressure because the ink is forced through the screen openings on to the final surface; screen printing can achieve a higher weight of transferred ink than the other methods.

Rotary and flat-bed printing

Early printing techniques – hand-carved wood blocks, movable type of wood or metal, lithographic stone – imposed the method of working in the flat; block or type was mounted in the press, inked, and brought into contact with the paper under more or less equal pressure over the whole surface area involved. As presses became larger to accommodate more work at any one time, so they became clumsier while adhering to this method. The mechanics of moving larger and heavier pieces of the machine in reciprocating or 'to and fro' motion imposed increasing strain on the design and power-efficiency of the equipment. Methods of printing from a cylindrical surface were therefore evolved so that the to and fro motion could be replaced by rotary action. The rotary method is now used for all high-speed work whether small offset equipment for the office or in newspaper and magazine production. Here again photo-methods of producing the printed surface have contributed to the ease with which rotary printing can be adopted.

The paper feed can be based on the use of individual sheets or continuous reels. Hence, sheet-fed and web-fed. Output is one of the deciding factors in this choice and for maximum production the combination of rotary printing with web feed is obviously attractive. Where quality considerations are paramount, other arrangements may well be preferable.

Connections between rotary printing and photo-processes

Whenever the printing is to be done from a cylinder rather than from a flat arrangement, problems arise in preparing the printing surface in its required design.

Rotary letterpress became possible with the invention of a process for making a curved mould from the original flat forme, which could contain both typeset matter and engraved blocks. Lithographic work was emancipated when thin metal sheets replaced the original stone and, being flexible, could be fastened round the cylinder of a rotary press. Photoengraving contributed to rotary letterpress by making it possible to etch a curved plate or to 'bend' a suitable etched flat plate.

Photo methods also made letterpress independent of actual type and its limitations. Gravure found it necessary to etch the copper cylinder required for rotary printing and evolved a special technique for dealing with this requirement. In this the photo-image is created on a sheet of paper carrying the photo-sensitive layer and referred to as 'tissue'. After imaging, the layer is transferred to the gravure cylinder to form the etching resist. A similar 'tissue' technique is used for creating the stencil which is then transferred to the screen for screen-printing.

All of these applications have made use of the photoreproduction systems based on chromium compounds; it is probably no exaggeration to say that the advancement of printing technology has depended on photochemistry as much as on any other single factor.

CHAPTER 5

PHOTOREPRODUCTION METHODS AND MATERIALS

For everyday practical purposes it is not difficult to distinguish the field of reprography from that of photography. The latter is mainly concerned with the production of a photograph of an actual scene or article, in colour or in black and white, and requires an apparatus of which the familiar camera is one example; the telescope and microscope are others. The production of portrait, pictorial, medical and scientific photographs is outside the scope of reprography. On the other hand, the use of a camera to photograph a drawing, manuscript or typed original for the purpose of making facsimile copies, could be a reprographic activity.

However, photographic and many reprographic methods function on similar principles and photochemical reactions (some reprographic exceptions appear in the Appendix). This being so, we will consider methods in terms of principles involved and not – as is perhaps more usual – with respect to the use to which the process is normally put.

Of the photoreproduction methods, those depending on silver compounds must take pride of place as to variety, flexibility and general utility. Materials sensitized with silver compounds have dominated the scene from the early days, although other processes are as venerable and still have limited uses today. Because of this domination, it has become customary to speak of silver and non-silver systems. The term unconventional is sometimes used, by which is simply meant processes not using silver.

We shall consider silver processes first, but it should not be thought that silver was the first substance to yield a photo-reproduction process. That honour belongs to the process invented by the Frenchman, J. Nicephore Niépce, and relying on the natural material called asphalt or bitumen. As a process this falls into the photo-polymer group (see pages 177–187); but it is proper to give Niépce his due place as the inventor of the first photographic process (1818–1826).

The two steps of photochemical change

Photo-reproduction processes achieve their results by inter-action between radiant energy and substances which are capable of absorbing that energy. We have seen that radiant energy results from exciting the atoms or molecules of suitable materials; that same radiant energy can in turn be absorbed by molecules of other substances which then undergo changes as a result of receiving this additional energy. These changes may be transient or permanent; they may be changes of chemical composition or merely into a more active condition. Whatever they are, before a useful photoreproduction system is obtained, these changes must be utilized to bring about a visual difference; that is, whether the radiation involved is within the visible spectrum or not, the reproduction process must conclude with a visible result: such as the production of a contrasting colour – a change from black to white, or vice versa.

Early workers in this field had of course no fundamental concepts on which to proceed. They worked by having some clear end-point in mind and pressing toward their objective, seeking and trying any idea which might help. Over the years, one by one, the basic ideas emerged. We should remember that much of what we take for granted had to be discovered by persistent experiment in the face of many failures. One such discovery lay in the early recognition that a successful method comprised two operations – first exposure of material to light or, as we can say, radiation; and second the steps whereby the effect of the exposure was made visible and/or permanent. This latter stage we now refer to as processing, which can be of several kinds, involving physical or chemical processes as necessary. The relationship between exposure and processing is important, as will be seen in the later descriptions, and more especially so in the case of silver-sensitized materials.

Origins of photography

The origins of the main photoreproduction methods lie in the first half of the nineteenth century. Some of the details are well authenticated and supported by sound evidence; some are being brought to light by searching through old notebooks and documents; some will probably never be unravelled and must remain speculative. The name of Niépce has already been mentioned. Other names which stand out from this period are Wedgwood, Daguerre, Herschel, Talbot and Hunt. Most of these saw the possibility of photoreproduction as a means to an end. Thomas Wedgwood's chief interest was probably in improving the application of patterns to the family's output of pottery; Niépce was interested in lithographic printing, and sought ways of producing printing surfaces without requiring the services of an artist; Herschel was interested in studying the sun's spectrum and found that chemical reactions would take his enquiries beyond either end of the visible spectrum. These pioneers contributed to photoprocesses, but photography as such was the clear endeavour of Daguerre and Talbot, whose work receives first mention in the account of the silver compound reactions which gave rise to photography.

Although the silver-sensitized materials gained a rapid ascendency in photography, other processes relying on salts of the metals of iron and chromium were invented and have survived. Today it is possible to propose other types of photochemical reaction; much time and money is being spent on examining new photo-imaging systems. Comparatively few new processes have 'come to stay', relative to those which emerged in the middle years of the 19th century.

Construction of photosensitive materials

In practice, it is not sufficient to find chemical substances which are capable of undergoing the requisite changes. They also have to be capable of incorporation into a photoreproduction *material*, usually in a sheet form which can be handled without damage and will survive any chemical processing baths which the reactions demand. We may say that a photoreproduction material brings the photosensitive substances into a position or condition to receive and react to radiant energy. Few photosensitive substances are self-supporting: they have to be carried on a base or support sheet. Thus the rudimentary reprographic product is in the form of a base sheet and a 'surfacing' of photo-sensitive materials. A third component of many materials is a 'binding' substance which serves to anchor the photo-sensitive materials to the surface of the base sheet and to keep them in position during the processing and in the final copy.

Paper as a base material

The base or support is in itself an important part of the material; it must combine strength when wet or dry, with ability to resist processing chemicals, and purity so as not to interfere with the photo-reproductive reactions. From the beginning, paper has been one of the most important supports in photo-reproduction. Today, papers are made for specific reprographic processes and to very close requirements of purity, strength, colour, etc. The early workers had to select from what was available, usually a writing paper of some kind. When the attempt was made to attach the necessary chemical substances to paper the first problem was to prevent the chemicals penetrating deeply into the paper, where they were hidden from incident radiation and inaccessible to further chemicals for processing. Therefore the majority of early processes had to commence by preparing the paper surface by a 'sizing' treatment. The earliest use of this term is in the 15th century in connection with the preparation of paper or parchment for illuminated manuscripts; reprography retains a connection with these 'arts' even into the 20th century. In the 16th century 'size' was a glutinous substance mixed with colours (pigments) for application to paper and cloth; by the early 17th century the term related specifically to gelatin.

The technologies of paper making, including sizing, and of gelatin in relation to silver-sensitized materials, have been mixed inextricably with the progress and evolution of photo-reproduction processes up to the present day.

Types of paper

All the papers normally encountered, whether for newspapers, books, magazines, writing, draughting, tracing, or other purposes, are made from naturally-occurring fibrous material. These are the cellulose fibres of plants and trees. Examples are:

> cotton (seed hairs)
> manilla (leaf fibres)
> straw, esparto (grass fibres)
> wood (softwood and hardwood fibres).

Papers frequently contain more than one type of fibre, careful choice being made to give the requisite qualities. In each case, the cellulose fibre has to be 'extracted' from the natural material and must be freed from undesirable matter which would interfere with its ultimate use. It must also be freed from foreign bodies which would give hard or dark specks in the finished sheet.

Most papers for photoreproduction require to be white and the cellulose 'pulp' is therefore given a chemical bleaching treatment; hence the terms 'bleached' and 'unbleached' as applied to pulp and therefore to the paper made from it. From the reprographic point of view, all processes involving chemical treatments are potentially dangerous because residual chemicals or by-products may interfere with the photosensitive substances which are to be applied. Thus it would be preferable to start with the purest possible clean, white pulp and to add the minimum of other substances; but a paper which is chemically 'ideal' may, in practice, be unusable: for example, filter-paper is pure, but is porous and very weak when wetted.

Paper-making processes

The paper is made by selecting the mixture of pulps or 'furnish' and subjecting it to a 'beating' process. The slurry of fibres in water is continuously 'beaten' by passing it between two rubbing surfaces which may be metal or stone. In the beating operation the fibres are not only intimately mixed, but may be shortened in length and also undergo varying degrees of 'hydration' (chemical combination with water). Other materials added to the beater contribute to the required final properties of the paper.

Eventually the finished pulp or slurry is flowed evenly on to a moving 'wire' – a continuous belt of wire gauze which allows the water to drain away from the fibres, so that they can begin to form the sheet. At the end of the wire belt further water is removed by squeezing between rollers or by suction, so that the wet web may have sufficient strength to pass on to the drying section. Even so, it must be supported until the drying process has achieved a sheet with reasonable strength, and therefore the first drying section of the paper-making machine has a continuous belt of absorbent 'felt' which carries the wet web round large drying cylinders. Finally the web becomes recognizable as paper, is dried down to the requisite moisture content and reeled up.

Some paper-making terms should be noted. Thus as the pulp slurry drains on the wire, the fibres take up a pattern referred to as the 'formation'. Formation may be uniform, or mottled, and the mottle 'large' or 'fine', these variations possibly giving fine-grain or coarse-grain effects in the reproduction material when sensitized. The side of the sheet in contact with the wire tends to retain the wire pattern even after the subsequent pressures between smooth surfaces; and forms the wire side, usually regarded as the back of the paper. The other side takes up a smoother, unpatterned appearance and is frequently referred to as the 'felt side'. As both sides are at some time in contact with the felt it is more precise to refer to the two sides as being wire side or non-wire side.

A further operation to which some papers are subjected is known as 'calendering', which consists of passing the web of finished paper through a 'stack' of heavy polished rollers, under pressure. The paper surfaces thereby acquire a higher (smoother) finish, and to some extent a more compact fibrous structure. Calendering does not materially affect the basic properties of the paper, which come from the nature of the constituent fibres, the beating process, any additives, and the machine operation.

Paper grain

Paper manufactured by continuous machine methods acquires properties associated with the direction of travel and flow at the point of formation, i.e. on the wire. This can be compared with the grain of a piece of wood and results from the tendency of the longer fibres to lie in the direction of travel (machine direction) as opposed to the across-the-web direction. The paper grain gives rise to such practical considerations as differences in tear-strength or ease in folding, and also influences speed of moisture re-gain on the edges of sheets and the temporary condition of 'wavy edges' in one direction. Sheets also have greater rigidity in the machine direction, which can be important in feeding through processing machines.

Paper-sizing processes

A paper sheet consisting of matted cellulose fibres and nothing else, is highly absorbent of liquids and has very little strength: when wetted it tends to disintegrate and revert back to a state of separate fibres. The fibres are therefore bonded together with a water-soluble but film-forming material (see page 96) by one or more of three methods:

1. Engine sizing, in which the necessary ingredients are added to the pulp slurry *before* the sheet is formed;
2. Surface sizing, which is carried out by applying the size to the surface of the partly-dried paper in a size press;
3. Tub sizing, which is an after-process applied to finished paper

by total immersion in the sizing solution, removing surplus and re-drying.

Apart from the economic differences, these methods achieve different effects on the properties of the final paper sheet. Engine sizing affects the whole body of the paper. Surface sizing affects the surface properties, offering considerable control as to depth and type of surface absorbency achieved. Tub sizing achieves varying degrees of penetration with effects somewhat intermediate between engine- and surface-sizing results.

From this brief description of the complex and highly developed processes of paper-making, it should be clear that the maker and user of reprographic products is very dependent upon the skill and care of the paper-maker. Much of the progress made in reprographic processes has in fact stemmed from the ability and the willingness of the paper industry to collaborate in producing paper to the necessary standard and, as far as possible, to a precise specification. As all papers depend on plant fibres, there is no absolute control of the basic raw material, which must vary with climatic and seasonal changes in rate of growth.

Transparent base materials: glass

While paper has played an important part in the evolution of photo-reproduction processes, the search began almost simultaneously for sheet material which could provide a more transparent support. Initially the paper print was made translucent by oiling or waxing treatments, *after* the image was complete, as a means of making further contact prints with short exposure times. The attractions of glass were not overlooked but it needed a layer of material which could attach itself to the glass, and contain or carry the photosensitive chemicals.

Cellulose – the same fibrous plant material as used for making paper – had been reacted with nitric acid in 1846 to produce nitro-cellulose or gun-cotton; a less fully nitrated cellulose was also known under the name pyroxylin. Pyroxylin dissolves easily in a mixture of alcohol and ether, the product being called collodion.

On evaporation of the solvents from a film of collodion on glass, a thin clear layer of the nitrated cellulose remains. Adhesion to glass is not good, but can be improved if the edges of the glass are treated lightly with a weak rubber solution. Collodion-coated glass came into very wide use for producing photographic negatives. The plates had to be exposed wet and developed immediately; hence the portable dark-rooms of the mid-19th century.

The wet collodion process was capable of magnificent quality in terms of resolution and tonal values. It remained in everyday use by photo-engraving firms into the 1930's and no doubt still has its adherents in some quarters where resolution is of paramount importance. Part of its 'charm' was the hazard that after processing the plate and giving it a final rinse under the tap, the collodion film might detach itself from the glass, roll up and disappear into the sink waste. If the plate survived all the hazards, it was, when dry, very tough, and various forms of after-work were possible.

With the wet collodion plate there was an advantage also in that the glass sheet – being of necessity selected for freedom from flaws – could be cleaned off and repeatedly re-used. This was consistent with the utilitarian 'do-it-yourself' outlook of early photographic operators, which only slowly gave way to the widespread use of factory-made products.

Transparent film bases

The nitro-cellulose of collodion is also the basic material for celluloid, one of the earliest plastics. The attractions of celluloid as a flexible, lightweight alternative to glass were obvious; its disadvantages are flammability and discoloration (yellowing) on prolonged exposure to daylight. As a base for photo-reproduction materials it was important until it was replaced by other cellulose derivatives, principally the acetates. The two types are generally distinguished as 'nitrate' and 'acetate' film, the latter also being described as 'non-flam' film. Acetate films will burn, but not very readily.

Nitro-cellulose can undergo spontaneous and rapid decomposition and has been banned for children's toys and amateur ciné film.

Progress in base materials

Availability of a range of paper, glass and film bases made the photo-reproduction processes adaptable to widely diversified end-uses. Progress since the 1875–1900 period has been by way of refinements rather than fundamental changes. Glass has increasingly given way to plastic film as a base material, though glass remains the first choice where the highest degree of dimensional accuracy is necessary. On the plastics side, great strides have been made. Acetate film continues in use in various forms (different degrees of acetylation for instance) but totally new synthetic materials are having varied degrees of success. Outstanding among these is 'polyester' film, which is related to Terylene and is more accurately known as polyethylene terephthalate. The polyesters have been shown by several years' widespread use to have a dimensional stability sufficient for exacting applications such as 4-colour reproduction involving large sizes.

Prepared papers

One other class of base material should be mentioned: translucent papers made from a frequently 'all-rag' furnish (i.e. consisting of cotton fibre rather than wood, etc.) and prepared by impregnation with suitable resins to increase transparency to ultra-violet and/or visual radiation. As usual, such improvement is sometimes associated with shortcomings of one sort or another.

Another hybrid type of base occurs when paper is given a surface layer of cellulose acetate. The result is a very thin plastic film – 0.0005 to 0.001in. thickness – supported on 'transparent' paper. This give some of the advantages of plastic film but at a lower overall cost. Similar surface treatment can be applied to tracing cloth.

Advancing technology in the manufacture and adaptation of synthetic films is progressively diminishing the price advantage of these more traditional methods of achieving special results. Sequential operations applied to naturally occurring raw materials become uneconomic if the cost of man-hours increases at a higher rate than the cost of raw materials.

Colloidal materials and 'film-formation'

Brief reference was made on page 93 to film-forming materials. So-called solutions of starch, gelatin, gums and of certain synthetics such as polyvinyl derivatives, are useful in the sizing of paper to reduce absorbency and therefore give a better performance for, say, writing paper. These substances do not dissolve in water but form dispersions or colloidal solutions, rather than true solutions such as those formed by salts.

Salts such as sodium chloride when in solution in water are in a dissociated condition. When the water evaporates, the salts re-associate and come out of the solution as separate particles, usually crystalline

– the size of the crystals depending on their rate of growth. Colloidal materials of the type mentioned do not so much dissolve in water as 'stretch' their molecular structure so that the water fills the spaces formed. This leads to the condition known as a 'gel', familiar in many kinds of prepared food. When a gel dries out, the substances which were in colloidal solution dry down and 'in one piece'. As distinct from the crystalline salts, colloids tend to form continuous films when the water of solution/dispersion has been evaporated away in the drying process.

In this way, for instance, gelatin as a paper size bonds the fibres together and fills up the spaces between them or bridges them with a film which prevents or delays water-penetration. These same colloidal materials may have an entirely different function in a reprographic material because they can enter into a special relationship with the photosensitive chemicals. Modern photography (with silver salts) rests on the relationship between silver compounds and the gelatin in which they are formed. Other materials make use of similar relationships and colloidal materials are therefore used in the two different capacities of bonding together paper fibres as well as forming part of the photo-chemical system.

CHAPTER 6

SENSITIVE SYSTEMS AND REPROGRAPHIC MATERIALS

Many chemicals are photosensitive. Some undergo obvious changes, such as in colour, after absorbing radiant energy, while others acquire a different type of chemical reactivity. The mere fact of such changes does not provide a photoreproduction process or material. The changes brought about must be visible and be made permanent. This usually requires that the photosensitive substance be associated with another, which though not in itself photosensitive, will enter into reaction with the first to achieve the desired final result. Such a combination amounts to a photosensitive system.

For example, it was known before 1800 that silver compounds were darkened by light, but it was not until a method of 'fixing' was discovered that a photographic process became possible about 1840. Similarly the light-sensitivity of diazo compounds was known in 1881 but a successful practical process was not devised until 1922.

All photo-systems must follow certain basic patterns. Assuming, for example, that substance A is photosensitive, exposure to the radiant energy to which it is sensitive must produce a new or changed substance, B. If the exposure takes place under a pattern or design – a leaf or piece of lace perhaps – an image of A is formed on the background of B. In the simplest case the image would be fixed by washing, on the assumption that either A or B (but not both) would be insoluble in water or a suitable solution. If this could not be done, then chemical reactions would be required to produce a permanent result. Washing might still be necessary to render the copy free from the risk of further change.

Thus, processes of varying degrees of complexity arise, the search being for the one which should give the desired result with the minimum of complications. All can be assessed by the ease or difficulty with which image formation and image fixation are achieved.

Irradiation and exposure time

Image formation by the molecular changes referred to above takes place only after sufficient exposure time has elapsed. This introduces the quantitative aspect of photochemical processes. Radiant energy has been seen to arise from molecular states in the material, metal or gas, etc., from which the energy radiates. Just as changes in molecular state can produce radiant energy, so also does absorption of radiant energy give rise to changes in molecular conditions. This is the basic situation of photochemistry and the fundamental mechanism of photo-reproduction systems.

If the energy is *forced* into a substance it may have little option but to increase its temperature or disperse the energy as radiation. But when a material is *exposed* to radiant energy, it may be selective in absorbing energy of particular wavelengths only. There can be no photochemical situation without absorption into the molecule of radiation which the molecules 'accept'. The word 'actinic' is used to express this meaning. It is applied to the radiation to which a particular system responds; thus, what is actinic for one system may be non-actinic for another. Radiation sensitivity is defined by spectral sensitivity curves in the same way that spectral emission curves are drawn for light sources.

Quantum size and actinic qualities

We have already made passing reference to the unit of radiant energy, the quantum (see page 25). By 'unit' we must understand the smallest amount of energy in this form which can be radiated, absorbed, etc. The *size* of these energy units depends on the wavelength of the radiation concerned; in other words, radiation of different wavelengths involves basic energy units of different size. This is why the reaction of chemicals to radiation depends upon wavelength: the energy units of the radiation must be of the size required to affect the particular chemical substance in question. Thus to be actinic to a given substance, radiant energy must be of the wavelengths which the substance absorbs, and have energy units of the correct size to affect the molecular structure of the substance.

To illustrate this question of quantum size it is useful to note the energy unit associated with the various colour bands of the visible spectrum and beyond. Using the visible regions as associated with wavelength, the energy of a single quantum or photon of radiation is as shown in the table below.

Photochemical effects occur throughout the wavelength range tabulated, but clearly the higher energies associated with the quanta of blue and violet light and of the ultra-violet regions are capable of greater molecular disturbances than those of the longer wavelengths. The spectral sensitivity curve of a photoreproduction material may be thought of as a graphical representation of the extent to which the system is disturbed by the higher and lower energies of the quanta absorbed.

Energy of radiation quanta

Colour	Wavelength nm	Energy *
Red	700	2.8
Yellow	580	3.4
Green	500	4.0
Blue	450	4.4
Violet	400	5.0
Ultra-violet	350	5.7
	300	6.6
	180	11.0

*The unit is the erg $\times 10^{-12}$. For comparison purposes, however, the actual unit is not important.

Measure of photoprocess efficiency

In practical reprography it becomes necessary to compare the results obtainable from one process with those for another, in terms of the energy required to achieve the end-result. A single quantum of radiant energy can bring about the activation or change of state of a single molecule; but very few processes operate at a 100 per cent use of available quanta. The term 'quantum efficiency' is used as an index of the photochemical change which any process offers. The quantum efficiency is merely the ratio between the number of molecules changed and the number of quanta absorbed in achieving the change. An efficiency of 1 is theoretically possible, but most processes operate at rather less. An efficiency of less than 1 may mean that some molecules have lost their acquired extra energy, before they could take part in the intended chemical reaction.

A photosensitive system may require not only a substance that undergoes some change when radiation is absorbed, but also a second substance that reacts with the first to fix a permanent visible image. Thus quantum efficiency may have little connection with the final result when seen. This introduces a different concept of process efficiency: the amount of visible effect which can be obtained from a given amount of radiant energy. The term used for such magnification of the initial effect is 'amplification'; two of the more obvious examples are in conventional silver photography and in electro-photography (see page 232).

Image amplification

In silver photography as operated in camera and projection printing (enlarging) only a very brief exposure is required to produce a latent image, and the *chemical* process of development is used to convert the invisible latent image to black metallic silver. In terms of energy, only one millionth of the total amount required to produce the silver image is provided by the exposure; the remainder comes from the chemical reactions which take place in the development process. In the case of electro-photography a latent image is formed as a pattern of invisible electric charges; these attract to the image areas a fine black resinous powder so that again a very considerable amplification is obtained, this time by a physical method.

In the ordinary way, a photoreproduction process is used to produce a single copy of an original; this is obvious in, for example, making a diazo copy from a tracing. If ten copies are required, ten operations are carried out and each is a self-contained photo-process. Some photoreproduction processes are aimed at the provision of a printing master. Such a master might be a stencil for screen printing, a plate for lithographic printing, a block for letterpress printing or an engraved plate for gravure printing. The actual preparation of these printing masters involves a photoreproduction method but the copies required are made by a mechanical operation in a printing press or machine. This group of processes is considered as photomechanical but the photoreproduction *method* is legitimately within the scope of reprography. However, in this type of situation the concepts of quantum efficiency and amplification tend to have less significance than in photoreproduction proper.

Background to practical processes

Any review of photoreproduction processes must commence at the early work with compounds of silver carried out in the opening years of the 19th century. After the year 1840 other photosensitive substances and systems came into use, and eventually the silver-based materials gave rise to photography in the modern restricted sense of camera usage and materials. It is tempting to regard photography as a special case of general reprography; but if so this is not to belittle the place and importance of modern silver photographic materials but to see them in perspective and in relationship to the early attempts to utilize photo-effects in a practical way.

The ability of silver compounds to darken had been known since the 16th century but it was not realized that the darkening was due to the action of light until a German physician, J. H. Schultze, carried out experiments in 1725. He poured solutions of silver nitrate on to chalk, thus precipitating silver carbonate. Schulze had no method of fixing his images.

Salts of silver are more or less affected by light, sometimes only when in the presence of organic matter. But the compounds on which photographic processes are based are the so-called halides: compounds formed between silver and three of the halogen elements chlorine, bromine and iodine, giving rise to silver chloride, bromide and iodide respectively. In normal use, the term photography is applied only to processes using one or more of the silver halides (for example, bromide papers, chlorobromide papers).

Silver chloride is immediately precipitated when a solution of a soluble silver compound such as silver nitrate is added to a solution of a soluble chloride such as sodium chloride. As precipitated, it is a white substance, turning violet and later brownish in light. Attempts to use this change were made by Thomas Wedgwood and Humphrey Davy in 1802; they can be said to have made the first photoreproduction material by preparing paper and leather impregnated with silver nitrate or coated with silver chloride. Wedgwood and Davy produced copies of pictures painted on glass but they, too, had no method of fixing the image. When the copy was exposed to light in handling the undarkened portions underwent the same change as had produced the copy and the image slowly ceased to be visible against the background.

The camera obscura

A camera obscura is, literally, a darkened room. An early version was a small building with a lens and rotating mirror, or prism, in the roof which projected a natural-colour image of the surrounding

102

countryside on to a table. Tourists paid for the privilege of standing round to admire the view. The image produced was of low intensity and the room had to be darkened to make it sufficiently visible.

Another version of the genuine camera obscura was the device used by would-be artists to produce an image by lens and mirror on the ground glass top of a small box. It was the English scientist William Henry Fox Talbot who, in 1833, conceived the idea of fixing on paper the pictures produced by the lens of his portable camera obscura. He succeeded in 1834 and the world's first negative in the modern sense was made by him in 1835. Between 1835 and 1839 Talbot used his negatives to make contact positives by printing down on to a second sheet of the same sensitized paper.

Fixation and development

Talbot's fixation process relied on salt solution to remove the unchanged silver chloride, but it was imperfect and the results were not sufficiently permanent. He was, however, a close friend of Sir John F. W. Herschel, who was attempting to use chemical reactions to explore the sun's spectrum in the regions beyond the violet and red ends of the visible spectrum. To Talbot's work Sir John Herschel contributed the fixation of silver-compound images by sodium thiosulphate – the hypo of almost every modern dark-room. The remaining discovery before modern photography was born came from the chemical process of development, and the tremendous effect of this in reducing camera exposure times from 1 hour or more, to some 30 seconds (i.e. amplification of some 120 times).

In the history of photography, Talbot is remembered as the originator of the first practically successful process after the Niépce bitumen process and the relatively short-lived but beautiful Daguerrotype. In addition to Herschel supplying the modern fixation step, the Rev. J. B. Reade is considered to have supplied the essential idea of development. Reade is known to have discovered that gallic acid (extracted from galls) had an accelerating action on the speed of silver-halide papers, and he worked independently of Talbot on the fixation problem. He did not publish his results and information is therefore fragmentary.

Talbot's notebook of 1839 referred to Reade's work and in 1840 Talbot recognized the function of gallic acid as a developing agent for a latent (invisible) image.

In addition to the chemical contributions of Herschel and Reade to the Talbotype process – which Talbot patented in 1841 – we owe to Herschel the terminology used today. He coined the word photography in place of 'photogenic drawing' previously in general use; and the terms 'negative' and 'positive' to distinguish between the two stages of the new process.

The Daguerre process produced a positive result on a sheet of copper. The copper was plated with silver and brought into reaction with iodine vapour (iodine is a purplish solid, easily vaporized; the 'iodine' of the first aid box is a solution of iodine and potassium iodide in water and alcohol). Thus a surface film of silver iodide formed on the silver-plated copper. Sensitivity was low and very long exposures necessary, until a chance observation that a low-density image was much improved overnight through contact with mercury vapour. Thus Daguerre, like Talbot, discovered the ability to form a latent or barely visible image, and convert this by processing to the fully visible result. The Daguerre plate also required a fixation step. Its image was a matt white, visible as a positive when the plate was viewed at the correct angle (to avoid reflection from the silver background).

Talbot christened his process Calotype (Herschel referred to Kalotype) but his achievement soon earned the recognition of Talbotype. The Daguerrotype was announced on January 5, 1839. Talbot made his announcement on the 25th of the same month. Daguerre, with government backing and a state pension, established a flourishing portrait business. Talbot directed his process to architecture, pictorial work and even photomicrography. Had it not been for the use of chemical development, Talbot's work would not have survived. Both processes adopted the technique of producing the light-sensitive substances directly in or on the support. Talbot first soaked paper in silver nitrate solution, then dipped it into sodium chloride solution. This precipitated silver choloride among the paper fibres. The same technique became standard practice in other processes and was still used 70 years later in the infancy of the diazo processes.

Printing without development

We can now see that it was the processing of the light-formed image that created photography in the sense of camera-speed materials, and the general dark-room work of today using bromide-type printing papers. Silver-sensitized papers will darken in daylight without chemical development; these were the print-out papers which amateur photographers were still using until the 1920's. Exposure was to direct sunlight, judging the time by opening the back of the frame to inspect the result. Toning and fixing baths had to be used, and washing, for permanence. POP should perhaps be remembered as more nearly a reprographic than a photographic process: which illustrates how narrow a dividing-line exists at the basic level, whatever the differences in usage and result with more sophisticated materials.

Summary of development

The years 1800–1840 represent a most important period because it was in this relatively short time that the basic approach to photo-reproduction processes was established for nearly all subsequent workers. In the beginning the thoughts and experiments were inextricably mixed as to the requirements of contact printing as opposed to the camera record. It is useful to review these years for their eventual bearing on the reprographic processes:

1801 – J. W. Ritter: silver chloride blackened more rapidly by violet light.
1802 – Wedgwood and Davy: contact prints from drawings on glass.
1818–1825 – J. Nicéphore Niépce: attempts at camera recording; bitumen process.
1834 – Talbot: camera images on paper, imperfectly fixed; the first negatives.
1835–1839 – Talbot: positive contact prints from paper negatives.
1837 – Rev. J. B. Reade: photomicrographs on silver chloride paper moistened with gallic acid.
5th January 1839 – Daguerre: camera images (positive) on polished silver plates.
25th January 1839 – Talbot announced 'photogenic drawing'.
1st February 1839 – Sir John Herschel: fixation of silver chloride prints with 'hypo' (probably used also by Rev. J. B. Reade).
March 1839 – Talbot: higher sensitivity of silver bromide.
April 1839 – Talbot: speed increased by incorporating gallic acid into the paper.
September 1840 – Talbot: Calotype process of brief exposure, chemical development and fixation.

Basic reproduction requirements

From this point onwards there is evident a considerable cleavage

105

between the *contact* reproduction processes and the camera processes; but against this common background there had appeared the requirements for both:

1. Colour change on exposure
2. Fixation, to render the photographs unaffected by further exposure
3. Contact printing whether from drawings or from photo-negatives
4. Inclusion of chemicals for subsequent reactions, in the light-sensitive layer
5. Use of chemical development processes to utilize latent image.

Silver-contact materials

In materials for contact printing, the silver halides are invariably used today as 'latent-image plus development' products. Their sensitivity is mainly in the blue and violet of the visible spectrum and into the ultra-violet region, the peak responses being 248nm for silver chloride, 319nm for silver bromide and 438nm for silver iodide.

There are many attractions in silver-contact materials: high image densities are obtained for low illumination conditions; the image, being metallic silver, is usually a good black colour and if properly handled, of great permanence. Disadvantages are that wet-processing is necessary, sometimes with washing and drying but not necessarily so. Unfortunately cost is high and continues to rise owing to the growing cost of silver.

The Collodion wet-plate and origins of dry plates

The search for more acceptable negatives for camera use led to improvements in other directions. The original paper negatives suffered by introducing into the image, the pattern or structure of the paper support. This could be partially overcome by waxing but the image had already acquired something of the paper's structure pattern on account of the penetration of the sensitizing solution when applied.

The use of glass in the camera presented difficulties in anchoring the sensitive material to its smooth surface. Starch paste and albumen (egg white) were both tried by Niépce de St Victor – nephew of the older Niépce – about 1847 and led to the Niepcotype process. In this the glass plate was coated with albumen containing potassium iodide; by immersion in silver nitrate solution, a layer of silver iodide on and in the albumen layer was obtained. Thus the method of preparation tended to follow the technique of Talbot.

Collodion was suggested in 1850 by Gustav le Gray and led to the process of Scott Archer, 1851. Here the collodion coating contained an iodide or bromide or a mixture and was sensitized by immersion in silver nitrate, thus precipitating the silver halides in the collodion film. The plate had to be exposed in the camera while wet, that is, carrying a surface film of silver nitrate solution. This was essential to give the system the required sensitivity, and meant preparation, exposure and processing 'on the spot'.

The collodion wet-plate had an extremely thin film, which accounts for its very high resolution; if it dries, crystallization of its surface layer of silver nitrate solution destroys the collodion layer; if washed before drying to prevent this, it loses sensitivity. The search began for a 'dry' plate which could be prepared in advance, kept until required for use, and brought back to the dark-room for processing.

Russell's collodion dry-plate

In 1861, a Colonel Russell found it possible to make dry collodion plates by washing away the excess silver nitrate, then dipping in a 'preservative' solution before drying. The idea behind the choice of preservative material may have been that of a protective film like varnish, but the successful materials used were all organic compounds which could act as *sensitizers* to the silver halides present.

Influences on sensitivity

In the original sensitized papers of Talbot, sensitivity was obtained by having excess silver nitrate but also (fortuitously) the organic material included in the paper-making process. The switch to glass as a support to the collodion layer carrying the silver compounds removed from the total system the organic material; but the effect of the preservative applied to a washed collodion plate was to restore sensitivity. This exemplifies the dependence of mere photosensitivity on the presence of other materials before a photoreproductive *system* is achieved.

Although the people mentioned had achieved the transition from Talbot's paper negatives to a practical dry plate process in the short space of some twelve years, they also learnt by accurate observation, that the dry system worked better if it retained some slight degree of moisture. It must be appreciated that many materials which are dry so far as handling or touch is concerned, do in fact contain a certain amount of water: paper for instance commonly contains 6–7 per cent while being perfectly dry as far as can be judged by handling it. The importance of moisture when present in the collodion layer of the dry plate was in maintaining all components of the system in intimate contact and providing the medium in which chemical reactions could take place. Water is necessary to a large number of chemical reactions and by removing it reactions can be slowed down to the point of ceasing altogether when conditions are anhydrous.

Evolution of gelatin emulsions

While camera photography had made great strides with the invention of wet- and dry-collodion plates, the process was made difficult to operate because of the variability of the layers produced. The search for improvements produced attempts to form the silver halides as an emulsion in a viscous liquid which could then be coated or cast on to the plate. At first collodion was used and in 1865 Wharton Simpson described a 'collodio-chloride emulsion'. This departure from the Talbot technique of precipitating the silver halides actually in the layer of the material, gave greater control of the chemical purity and therefore greater consistency. The next major landmark, however, is 1871 when Dr R. L. Maddox gave practical details of the preparation of 'gelatino-bromide emulsion' and of 'dry plates'.

Gelatin has been referred to earlier as a colloid and as having the property of forming films when a 'solution' or dispersion of it in water dries. One attraction of gelatin over collodion was in the ability to work in water instead of in the alcohol/ether mixture in which pyroxylin has to be dissolved. Gelatin placed in cold water merely swells by absorption of water into its mass. When it is fully swollen slight heating brings about complete dispersion of the gelatin. On cooling the whole forms a gel, and during cooling the dispersion is viscous. In this condition a gelatin dispersion will hold solid matter in suspension, that is, prevent it falling to the bottom of the vessel. This property of gelatin has made possible the photographic emulsion of modern materials, and thus Dr. Maddox may be said to have brought all the earlier work to the point at which silver photography had 'arrived', almost a century ago.

To prepare the emulsion, gelatin is dispersed in water and one of the reacting chemicals dissolved in the dispersion. The other chemical, dissolved also in water, is then added while the dispersion is rapidly stirred. Thus silver nitrate might be in the gelatin dispersion, and a solution of the appropriate halide added to it. The quantities of silver nitrate and halide salt are carefully regulated so that no excess silver nitrate remains; it is all precipitated as silver halide, in a very fine state of sub-division. On cooling, the mixture gels, making it possible to break it up and wash away all the soluble salts leaving the pure silver halide suspended in the gel. When required, the emulsion merely needs warming to bring it to a liquid condition in which it can be coated on glass, film or paper.

Ripening of gelatin emulsions

As described, the emulsion is not particularly sensitive to light. But it soon became noticed that sensitivity increased on standing – say for a week at the temperatures of a warm room – and that this 'ripening' could be accelerated by heating. The ripening process is extremely important in modern emulsion making, as also is the influence of the gelatin itself. Different types of gelatin and different batches of the same type, show different effects, so that it becomes necessary to use careful blends of gelatin in order to achieve consistent results.

The influence of gelatin on the properties of the final product shows that in silver-salt photography the gelatin not only serves as the medium in which the photosensitive substance is held, but also enters into the system and takes a part in its chemistry. Collodion in contrast is a convenient but inert substance which serves only as a physical means of creating the thin photosensitive layer required.

Dye sensitization of gelatin emulsions

Ripening increases sensitivity but does not effect any fundamental change in the *character* of the response to light. It is, however, possible to change the behaviour of a silver halide layer so that it responds to a wider region of the spectrum. This is done by the addition of certain dyestuffs which are therefore called spectral sensitizers. A change in spectral response need not necessarily mean a change in speed and the two types of sensitivity modification should not be confused. The function of a spectral sensitizer is to absorb radiation and pass on its acquired energy to the sensitive substance, repeating this process without itself being changed structurally.

Unless they have been spectrally sensitized, silver halide materials are sensitive only to the blue and violet parts of the visible spectrum and some way into the ultra-violet region. In this form they are referred to as 'ordinary' emulsions. By increasing the response to include the green and part of the yellow regions, so-called ortho-chromatic emulsions are obtained. Further extension of the response may be obtained to include the red, in which case a panchromatic material results. In each case the original blue/UV response remains, i.e. the spectral response is extended additively.

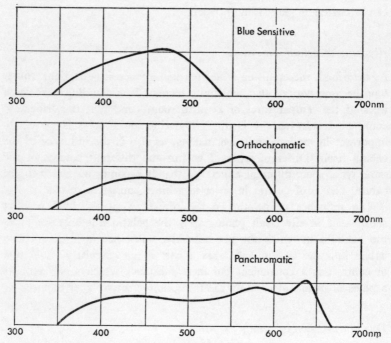

Spectral response of silver halide emulsions.

Safe light requirements

A panchromatic emulsion thus responds to the major part of the visible spectrum. Of all the reprographic processes, this type comes nearest in its response to what the eye 'sees'. To this extent it comes closest to the theoretical ideal for reprographic purposes; but in practical terms it would not be acceptable in the majority of cases because of expense and the demand for dark-room handling and processing.

For the ordinary emulsion, the actinic region ends at a little over 500nm. Thus light of some 550nm upwards is 'safe' for handling this type, being non-actinic. Such light would be quite unsafe for handling the orthochromatic type. Similarly, light of 600nm upwards which might be safe for the orthochromatic emulsion would be unsafe for panchromatic which demands total darkness for safety against accidental exposure or 'fogging'.

Principles of stabilization methods

The photographic papers of reprographic interest are gelatin/silver-halide systems coated upon a base paper. They rely on brief exposure to give a latent image, which then requires the chemical process of development to make it visible. A further fixation process is necessary to convert residual silver halide to soluble salts which can be removed by washing so as to give permanence. As an alternative to the separate processes of development, fixing and washing, greater convenience is available from so-called stabilization processing.

In stabilization materials, the clock is put back to the early Talbot papers in which the developing agent was included in the photosensitive layer. Modern materials ensure that the chemical process of development cannot commence until the layer has been made alkaline. Therefore the exposed paper is 'activated' by immersion in a simple alkaline solution. Development of the image is almost instantaneous, the print then passing into a second bath which stabilizes the print. In this step the unused silver halide which ordinarily is removed by fixation and washing, is converted into silver salts which are no longer light-sensitive (in the sense that the print can after stabilization be freely exposed to light without any darkening of the white parts).

Baryta-coated paper

Silver halide print materials tend to be used in the main where their expense is justified by the qualities required in the finished work. When delicate tonal values are involved, it would be intolerable to have the pattern of the base paper showing through the true image on account of the emulsion finishing up in different thicknesses. Therefore the emulsion must be coated on a very smooth or 'level' surface. This cannot be achieved in the paper-making machine so an artificial surface is created by what is called 'baryta coating'.

For this purpose, a suspension of very fine barium sulphate is prepared, again in gelatin, and applied to the photographic base paper

111

in one or more layers according to the degree of finish required. The gelatin is hardened with, for instance, chrome alum, so that it will not re-melt when the photographic emulsion is applied. The photographic base paper is itself of a very high quality. It must be chemically pure because, just as the silver halide system is influenced beneficially by trace quantities of, say, a spectral sensitizer, it is also prone to suffer from minute quantities of desensitizing substances or other interfering chemicals.

CHAPTER 7

REPROGRAPHIC USES OF
SILVER HALIDE PROCESSES

Talbot's contribution to photography lay in devising a negative/
positive two-stage process, giving the ability to make the negative by
a single camera operation and any number of positive copies by
contact printing. Silver-sensitized materials are, however, inherently
expensive and for reprographic work the production of a single copy
might involve both an expensive negative material and an equally
expensive positive sheet. Although the cost may be reduced by making
several copies from one negative, it remains high relative to other
processes. For some time the photostat system of paper-negative single
copies and from those, paper positive copies was the principal method
of copying books and opaque originals.

In photography, silver-material costs are justified by the specific
qualities of the final result. In simple reprographic work, silver is
justified by requiring less elaborate equipment – therefore lower
initial outlay – and by giving operator convenience and reliability.

Diffusion-transfer systems

One of the more successful document-copying processes relying on
silver-sensitized material is the so-called 'diffusion-transfer' system.
The Land Polaroid camera, in which chemical processing after
exposure is carried out by breaking 'pods' of solution within the
camera back, uses the same principle. It is an interesting thought that
this principle was rediscovered more or less simultaneously by three
people working independently in three countries. These were Rott in
Belgium (Gevaert), Weyde in Germany (Agfa) and Land in the USA
(Polaroid). The earliest description of a transfer-type of process came
in 1857 by Lefevre, other observations at intervals in the 1890's and
the 1920's. The exact history of the modern diffusion-transfer process
was confused by the second World War years, but has been well

113

recorded since then; see for instance Chapter 16 of Neblette: *Photography: its materials and processes* (5th Edn. 1952. MacMillan).

Diffusion transfer uses two sheets of material – one to give a negative and the other, the positive; but only the negative sheet carries expensive silver-sensitizing salts. Exposure is usually by reflex methods (see page 210) but not necessarily so if suitable single-sided originals are available. The negative sheet also includes its developing agent so that development can take place by merely immersing the exposed sheet in an alkaline bath. This bath contains additionally a fixing agent, so that the bath is capable of converting silver halide to water-soluble silver salts, as development takes place.

The positive sheet of diffusion-transfer contains no silver salts, but a gelatin layer with development nuclei dispersed in it. The function of the nuclei is to precipitate metallic silver from a solution of silver salts, without any involvement of light; the precipitation is a purely chemical reaction which takes place as soon as the reactants are in contact.

In use, the diffusion-transfer negative sheet is exposed and then put face-to-face with a positive sheet and the two passed together through a processing machine with the alkaline-fixative bath described. As the sheets enter the bath (travelling below the surface) development of the negative immediately commences; at the same time the fixation reaction can start but is less immediate. The two sheets then emerge from the bath and are squeezed between rollers into intimate contact. Surplus solution is returned to the bath and the sheets emerge 'moist' rather than wet.

There is, however, enough solution retained between the two sheets for diffusion transfer to take place. Wherever the negative sheet was *not* affected by the exposure, i.e. directly over the black parts of the original, the silver halide is not developed to metallic silver but is taken into solution by the fixing agent. When in solution, it is then free to migrate from its original position, and diffuse into the gelatin layer of the contacting positive sheet. Here it meets the development nuclei and is promptly precipitated as metallic silver. After a sufficient time in contact, the two sheets are peeled apart. The positive copy is allowed to dry, the negative is discarded but may be given normal fixation and washing if required as a possible source of conventional contact prints at some future date (its image is laterally reversed and unsuitable for reading as a copy). A good DT negative can give satisfactory diazo prints which are also negatives.

Apart from the convenience aspects – equipment is simple and easily maintained – the silver salts of the one negative sheet are fully utilized: those parts which do not produce the negative image transfer to the positive sheet and there provide the positive image.

The diffusion-transfer type of photocopying had a tremendous vogue during the 1950's, but has given ground more recently to the electrophotographic systems (see page 232). It still forms a very reliable system, attractive where initial cost of installation must be small but cost-per-copy is not a prime consideration. Variants of the process as described are double-sided positive sheets which are processed as a sandwich between two exposed negative sheets; and a film positive sheet which can then be used as a master for subsequent copies, for instance by a diazo-type process or by lithography.

Diffusion-transfer can also be used to generate a metal lithographic printing master in the one operation: instead of a positive sheet carrying development nuclei the aluminium master is processed in contact with the negative. Silver salts unaffected by exposure and development are again dissolved by the bath and migrate to the aluminium sheet where they deposit metallic silver in a tenacious form which is very suitable for offset-lithographic printing (mainly at the office-duplicating level).

Negative papers for diffusion-transfer are made in a variety of speeds such as low for handling in normal room lighting or high-speed for projection printing and requiring dark-room conditions in processing. Positive papers are in various weights, air-mail to single card thickness, or can be made transparent as a low-cost alternative to film.

Direct-positive materials

Another group of silver-sensitized materials of increasing importance in reprography are referred to as 'auto-positive'. These omit a negative stage altogether, relying on the so-called Herschel effect. Herschel noticed that a latent image in a silver-sensitized material could be reversed to a positive by a second exposure. From this 'effect' it has been possible to manufacture 'pre-fogged' materials which if taken straight from the packet and developed, would become black all over. But by giving the pre-fogged sheet a second exposure to an image the 'fog' is lost where light acts and only remains where there is no light action. In practice, auto-positive emulsions are 'fogged' by chemical treatment so that when developed without exposure, the emulsion blackens to its maximum density. The effect of exposure is to *prevent* development, so that the layer is positive-working. As this is the reverse of normal photographic operations, the material is often called reversal material but in diazo-processes, positive-working is normal and the term reversal is used for a negative-working system (see also lateral reversal, page 221).

Auto-positive or reversal silver materials are useful as both papers and films for normal contact-printing; there is also an excellent film product for reflex working and producing an image density very satisfactory for lithographic plate-making. Another direct-positive material is known under the description 'wash-off'. In this material – based on polyester film – the emulsion is constructed so that during development the oxidation products of the developing agent harden, or 'tan', the gelatin of the emulsion. This tanning action must be restricted to the actual image areas, and there must be no appreciable diffusion of oxidation products sideways into the adjacent areas. After development the unexposed parts are completely washed away with warm water, so that the final print comprises the base film and the image parts only of the original gelatin layer. Thus the copy retains the desirable properties of the polyester base more fully than if it still carried a complete coating of gelatin (the dimensional stability of polyester to changes of atmospheric humidity is adversely affected by gelatin, which contracts in dry atmospheres.)

An office-copying system (Eastman Kodak's Verifax) also using the hardening principle is gelatin dye-transfer. The gelatin layer of the silver halide emulsion is dyed. Development of the exposed sheet by activating bath results in hardening of the exposed areas. By pressing the developed sheet on to a suitable copy paper, then separating, some of the unhardened dyed gelatin transfers to the copy sheet. In this way, positive copies are obtained – the hardened negative image remaining on its original support. Materials for single or multiple copies are available, but copies are progressively weaker as the dyed layer is progressively transferred. The transfer takes place under moist conditions, which again imposes a limitation on the number of copies which can be taken before the negative sheet becomes too soft for further use.

This dye-transfer system can also be used to prepare a lithographic master, but a more elegant method is the 'photo-direct' system of Kodak. This provides a camera-speed material, allowing direct exposure of the master sheet to the original material on the camera copy-board. Thus, enlargement or reduction is available, and without making a conventional negative for printing-down to the plate. Automated processing further simplifies the operation.

The master-sheet for this process carries three layers: lowest, or next to the base, is a layer of the photographic developing agent; next, a normal silver-halide emulsion; and at the top, a silver-halide layer of pre-fogged, auto-positive type. Light from the white parts of the original (i.e. non-image) passes through the top layer causing photographic reversal. It continues on to the middle layer, causing normal exposure of that layer. In the activating solution, this middle layer develops where affected by light, because of upward diffusion of developing agent from the lowest layer. Where the middle layer was affected by light it can therefore develop, but the developing agent is thereby oxidised and in those areas becomes exhausted before there is any effect on the topmost layer.

In the image areas – black in the original – there is no exposure of the middle layer, therefore the developing agent is free to migrate upwards and to reach the top layer. Hence a positive image develops in this top layer. As this image is developed, the gelatin of the top layer is tanned or hardened by the development oxidation products. Thus the image areas become ink-receptive, while the non-image areas remain water-receptive.

Miniaturization

Yet another major contribution of the silver-sensitized materials to reprography is in the field of micro-images. The original idea behind microfilming was to save storage space, but with the availability of mechanical sorting equipment and the proliferation of documents and drawings, miniaturization can also much improve *accessibility* of stored information. This is not the place in which to describe the complexities and potentialities of micro-records, storage and retrieval; sufficient for the moment to point out that the majority of such systems rely on and revolve around, an original micro-image on a silver-sensitized material.

Micro-records of normal documents may be on 8 mm, 16 mm or 35 mm width film stored usually in reels, but the largest also as single frames mounted in cards. Micro-records of engineering drawings commenced at half-plate ($6\frac{1}{2}$ x $4\frac{3}{4}$ in.) size but are now successful on 35 mm and 70 mm film. The last-mentioned is also satisfactory for such critical work as the 25 miles-to-the-inch Ordnance Survey sheets. In libraries – whether of books or typewritten reports, etc. – the micro-images may be carried as numerous tiny images on a sheet 6 x 4 in. (twelve columns of 6 frames each, or 72 pages per sheet; referred to as a micro-fiche).

Miniaturization systems can only succeed within the limits of quality of the original micro-negative on a silver-sensitized material. The requirement is that of sufficiently high resolution but working contrast is also of importance. Materials of this nature must be used with especial regard for exposure and processing control: manufacturer's instructions must be followed with care if results are to be satisfactory.

Contrast and tonal reproduction

In the ordinary course of conventional photography one expects what are called good tonal values. The process of photography with silver-sensitized materials is capable of reproducing a wide range of the tones, or dark to light shades, of a scene. In this way a rounded surface such as the cheeks in a portrait, is reproduced with a subtle change of tone from the highlights to the shadow areas. The quality of a photographic material which expresses its ability to reproduce tones is referred to as its contrast. If the photographic process resulted in too low a contrast, the result would look flat and lifeless. If too high a contrast, the result would be unpleasing and receive the descriptive comment, 'soot and whitewash'.

Reprography frequently requires a high contrast result. Copying a typewritten letter for instance, we like to get as black a reproduction as possible, with a white background; but there must be retention of the weaker parts of the typing. The danger in aiming at a high-contrast result lies in the tendency for middle tones to be lost; below a certain value they reproduce as whites, and above that value tend to merge into the maximum densities. Thus a strong black print on a clean white ground arouses suspicion as to what has been lost in the reproduction process.

The working contrast of a photographic layer derives from two things: first, the inherent properties of the silver emulsion; second, the nature of the chemical developing solution. Additionally, contrast can be increased by reducing exposure time and giving increased development time. Hence the importance of adhering closely to the manufacturer's instructions for processing. In the main, silver-sensitized products for reprography are of intrinsically high working contrast. This applies not only to 'direct' copying materials such as the diffusion-transfer and auto-reversal products mentioned, but also to the materials used in connection with the printing processes. (Photogravure is an exception – see page 138).

The latter type of silver-sensitized material is frequently referred to as a process material. By this, it is meant that the plate or film (usually for use in the process camera) is capable of giving a high-density image at high contrast and low background density. Good quality negatives of this type make all subsequent operations comparatively easy, whereas a negative of indifferent quality calls for much 'dodging' or manipulation to achieve a satisfactory result. In addition to the more traditional process materials, there are also lith-type films which by careful attention to development detail, give even higher densities with very clean backgrounds. As the name implies, these have been especially useful in the making of lithographic plates, but their value is not restricted to this; photo-typesetting on lith emulsions is as useful to photoengraving as to the lithographic plate-maker.

Reproducing continuous-tone images

Printing processes – other than photogravure – rely on high-contrast images, so we must consider how it is possible to reproduce by relief and planographic printing, the tonal values which are essential to the majority of printed illustration. To do so, we look first to the earliest of methods for reproducing illustrations, which before the invention of photography had to rely on hand methods of preparing the printing block. The techniques were to engrave lines on a metal sheet either directly with an engraving tool; or by coating the metal with wax, scribing the wax away in lines and chemically etching the metal where thus exposed. Combinations of hand- and chemical-engraving were also used.

The result when printed was a picture in which both the outlines of the various parts as well as the shading of light and dark areas, were produced by varying the size and frequency of *lines*. Rounded surfaces could be represented by lines of changing width; widest in the shadow areas, finest in the highlight areas. For deepest shading, cross-hatching was used. The visual effect of varying densities was obtained, though the lines themselves all printed at the same density.

This was the best that the printing process could do to match the work of the artist with pencil or colour, in which shading is obtained by the simple expedient of putting more or less graphite or pigment on the different areas. In these cases the gradation of one tone to another could be much more gradual than with the engraved printing plate. Such subtle gradations are referred to as *continuous tones* as distinct from the line work which could only simulate or imitate, tones.

Half-tone process

As soon as photography was invented, there was immediate interest in applying the new processes to the production of more subtle tonal gradation in printed material. The first such process came along as early as 1852 (Talbot's first photogravure method). But letterpress or relief surface printing had to await the invention of the crossed-line screen and the perfection of the wet collodion process before it could move away from simple line illustration. A positive photograph (black and white print) became the starting point of the photoengraver's half-tone printing-block. This positive print was re-photographed to make a new negative; but the new negative was made by exposing it through a screen mounted within the camera. The screens were made by ruling parallel lines on each of two glass sheets, then mounting the sheets face-to-face with the rulings at 90° to each other: hence, 'crossed-line' screens.

The effect of the screen on the new negative was to impose upon the image a dot structure. The blackest parts of the new negative had a small regular pattern of white dots; the lightest parts in the negative had small black dots; intermediate tones progressively approached a 'chess-board' pattern of precise squares with corners touching. Thus all the tonal gradations in the positive print became reversed in the negative in the sense of dark to light and vice versa; but additionally, tones of the positive were represented in the negative by patterns of varying dot size.

When this 'screened' negative is used for making the actual printing-plate, the dot pattern is transferred to the metal surface as an *etching-resist*, and suitable chemical solutions used to remove the metal between the dots. This treatment of the original photograph tended of course to lose the sharp outlines of objects in the photograph, but the finer the screen, the better definition was obtained and the more gradual the apparent tonal transitions.

Thus, 'line' illustrations and 'half-tone' illustrations came to be regarded as distinct and separate types of work. There is no real distinction, because in both cases the line or dot of the block is intended to transfer a full weight of printing ink. Both in half-tone as well as in line work, the *weight* of transferred ink is uniformly high. The tonal appearance in the result from the screened block is an optical illusion produced by varying dot sizes, and not by differing concentrations of transferred ink.

Producing screened negatives

There is a great deal of expertise in the production of good screened negatives in the camera. Techniques of varying complexity are required to get the best from any given original subject and the results depend to a great extent on the skill of the camera operator and that of the photoengraver or blockmaker. In recent years, simplified methods of screening have become available. These involve so-called *contact* screens. They are used to produce screened positives from continuous-tone negatives. The screen is used in contact with the positive material, either in the printing frame or on the enlarger base-board when printing by projection. These newer methods are especially attractive today, when so much reproduction is required in natural colour. Starting frequently from a colour transparency, continuous-tone separation negatives are first made by using filters to separate the colours into the basic amounts of magenta, cyan (blue-green) and yellow finally required (so-called, analysis filters).

Positive films are then made from each of the continuous-tone separation negatives, using contact screens to break the tonal grada-tions down to the appropriate pattern of dots. Where possible, these positives can then be printed straight down to printing-surfaces with positive-working sensitizers (for letterpress blocks or lithographic plates). If negative-working metal sensitizers are to be used, the screened positives must be re-contacted to convert them to screened negatives. Although this may seem a cumbersome procedure, these various photographic stages do allow some after-treatment to be carried out. An area may be locally treated with chemical solutions to reduce the dot size and suppress that particular colour, or to open-up a shadow dot so as to lighten the colour impression of that area.

Apart from the applications of screened images to the printing processes, it is also useful to print screened positives of normal photo-graphs on to diazo-sensitized materials, which are usually of too high a working contrast to produce good tonal gradations. As each dot of the screened image is required to reproduce at maximum density, it is an advantage to print on to a high-contrast paper as described on pages 154 and 155.

A final word on screened images: references will often be found to 'percentage' values. These refer to the total amount of any given area which in the final print, receives ink from the printing-plate – expressed of course as a percentage of that area. Thus the black dots in a highlight area might be 5 per cent, or the white dots in a shadow area 10 per cent, etc. The efficiency of a sequence of operations leading to the printed page may be expressed as the percentage highlight dot it can retain. Thus it is better to use a process retaining a highlight dot of 5 per cent size than one which cannot retain less than a 10 per cent dot.

CHAPTER 8

PROCESSES USING CHROMIUM COMPOUNDS

Important as the silver halides proved with their capacity for adaptability and refinement, many other potential systems were studied. In the process of natural selection, two which rose to importance and have persisted to the present day are reactions involving iron salts, mainly giving rise to the blueprint, and the group relying on compounds of chromium which have largely been responsible for the photomechanical processes.

The processes using chromium compounds were all necessarily negative-working. They came into use predominantly in photomechanical situations and have stayed in use for this purpose despite certain practical difficulties or limitations.

As with silver compounds, the fact of light-sensitivity in chromium compounds was known for some time before a primitive photoreproduction system emerged. Suckow recorded the fact in 1830, but it was 1839 before Ponton achieved a printing-paper.

Ponton dichromate process

Ponton impregnated paper in a solution of potassium dichromate and dried it in the dark. When exposed to light this paper turned brown. An object placed on the paper during the exposure had its outline reproduced by preventing this change taking place. The exposed paper only required washing in water to dissolve away unchanged dichromate, making the image permanent or 'fixed'. We now know that the brown negative image resulted from reaction between the dichromate and the sizing materials used in paper-making; that is, the photo-system comprised potassium dichromate *in conjunction with* organic material from the paper sizing.

An ingenious process followed Ponton's paper and can be considered the ancestor of nearly all subsequent processes relying on chromium compounds.

123

Becquerel's dichromate-starch process

The Becquerel system dates from 1840. A mixture of starch and dichromate was applied to the paper and dried. Exposure to light resulted in the starch losing its solubility in water. As with the Ponton paper, washing in water after exposure removed the dichromate/ starch from the unexposed areas. The print was then treated with a solution of iodine, which forms an intensely-coloured blue complex with starch.

Effect of 'development'

As with the silver-salt processes of Daguerre and Talbot, so also in the case of chromium compound systems like Becquerel's, there are the two distinct stages of exposure to the effective radiation, followed by chemical treatments of one sort or another. But the part played by chemical development of a latent silver-halide image differs from other processes in that the energy of the photographic development reaction contributes very largely to the total imaging effect (see page 101). It is this which gives the silver processes their unique position among photochemical systems as camera-use materials. With the dichromate system, exposure has to contribute the whole of the energy required for creation of the image; the after-processing is then used to make the image permanent and effective for the purpose in mind. In many cases the first processing stage, as with Becquerel's paper, is a water-wash to remove the unused and unwanted non-image parts of the original photosensitive layer. Washing completed, there remains a complete, permanent and visible image. The second stage, reaction with iodine, 'develops' the image in the sense that while the exposed parts owe their visibility to a brownish colour from the chromium compounds, the starch remaining in the image areas is now converted to the very dark blue starch-iodine complex.

Because of the dominance of the silver processes based on Talbot's method, it has become customary to use the term 'development' for almost any processing after exposure. This use of the word is so established that any attempt to introduce changes would confuse rather than clarify. While following the convention of 'developing' the images of processes other than silver-based, we should make a mental distinction between the chemical process which is essential to completion of the silver, photographic, latent image, as opposed to the treatment given to a completed image to make it permanent, or more visible, or suitable for some further purpose.

Talbot's 'photoglyphy' process

Following his first process, Becquerel also found that the action of light on other organic layers containing dichromate was to render them more difficult to re-dissolve in water. This effect was found with gelatin, casein, and albumen, for example. However, it was Fox Talbot who found a use for this discovery in 1852.

Steel or copper plates were coated with a gelatin/potassium dichromate solution. On exposure the light-struck parts became hardened and insoluble in water. By washing the exposed plate in warm water, the unexposed parts (protected by the original or object to be reproduced) were dissolved away leaving bare metal in these areas. This made it possible to attack the unprotected metal with suitable chemical solutions. The function of the colloid image was thus to protect the metal from chemical attack by *resisting* the solutions employed; hence the terms 'resist' or 'photo-resist', applied to this type of work.

In conjunction with the known processes of engraving metal plates, Talbot's 'photoglyphy' was the first photoengraving process of practical importance. The areas etched away chemically formed a recessed image; this accepted printing ink which could then be transferred to a sheet of paper by placing it in contact and exerting sufficient pressure to force the paper into the inky depressions. Talbot took the further important step of placing a gauze between object and sensitized metal. Thus the large open areas were broken down into small ink 'cells', introducing the principle of giving the copy tonal values; that is, ink densities intermediate between the maximum density of the full weight of ink and the minimum density achieved from the non-etched parts of the metal. Talbot's methods were modified by the Czech, Karel Klič and in that form remain in use in the photogravure process (see pages 138–140).

Nature of chromium compounds

Chromium compounds have given rise to a steady succession of photoprocesses not directly relevant to reprography. These use the hardening principle applied to a pigmented layer. For the reprographer, greater importance attaches to the use of etching resists for photomechanical purposes. Before describing these in greater detail some further explanation is necessary of the basic photosensitive system under which they operate. Such systems involve two essential components: the chromium compound and a colloidal substance.

Chromium metal is familiar as the hard polished deposit of chromium plating. It is capable of forming simple metal salts such as chromium chloride or chromium sulphate. The compounds used for photosystems are of a different type; in these the metal chromium behaves as if it were an 'acidic' substance. When functioning as an acid, chromium forms two series of salts known as chromates or dichromates. These merely differ in the amount of alkali in combination with the acid, in fact chromates and dichromates are easily converted from one to the other by adjusting the amount of alkaline substance present.

*Di*chromates are sometimes referred to as *bi*chromates; but chemists prefer to keep the latter term for what is known as an 'acid salt' – e.g. sodium bicarbonate.

The chromates most commonly encountered are those of sodium, potassium and ammonium; lead chromate is a bright yellow substance, found naturally or easily made, and used as the pigment 'chrome yellow'. Generally speaking the chromates are yellow in colour, but when converted to dichromates are usually a deep orange-red. The choice for sensitizing purposes lies usually between ammonium dichromate and potassium dichromate, with a preference for the former on account of higher solubility in water. This allows the use of higher concentrations and discourages harmful crystallization from occurring as the coating dries. Sodium dichromate is less easy to prepare in a pure form; it is used in tanning, hence the term 'chrome leather' usually indicating a very tough variety.

Choice of colloidal material

The colloidal materials for these systems can be very varied. In the main they are natural products and can be of three types:

Protein material:	gelatin, fish glue, casein, albumen, process glue.
Carbohydrates:	starch, gum arabic.
Resinous:	shellac.

Synthetic materials can also be used in conjunction with dichromates, predominantly polyvinyl alcohol, polyvinyl pyrrolidone, urea-formaldehyde resins and the like.

126

Speaking rather loosely, all these compounds are of a large-size molecular structure, with the ability to form a lattice or network by cross-linkages between molecules. In dilute 'solutions' or above certain temperatures, the substances probably exist as dispersions of separate large molecules moving freely about in the liquid. When concentration is increased or temperature reduced the separate molecules become tangled together in a network and groups of adjacent molecules may 'cross-link' to connect the whole together.

In the drying of a layer of, say, a simple gelatin solution, the water of the solution can at first evaporate freely but later on is 'trapped' in the colloid structure. If drying is rapid a surface layer or skin will form and prevent the easy escape of water from the lower layers, but with care the whole of the layer can be 'dried down' to form a continuous sheet or film (though it may still contain a few per cent of water which allows some flexibility; completely dry gelatin is hard and brittle). Such a colloid film may be re-dissolved in water by allowing it to swell, i.e. to re-absorb water into its mesh-like structure, and then raising the temperature to disentangle the molecules and break some of the linkages formed during drying.

However, certain chemicals will interfere with the process of re-solution. Examples are formaldehyde (a gas, but usually met as formalin, a solution of the gas in water) and salts such as chrome alum (a double salt of potassium sulphate and chromium sulphate). The by-products formed in the process of photographic developments are also sometimes capable of acting in this way, resulting in so-called 'tanning development' (see page 116).

The action of small amounts of these substances does not render gelatin completely insoluble; they make the gelatin film tougher in the wet condition and it becomes possible to work at higher temperatures than are normally safe, such as those used in rapid processing. Similar changes take place in the presence of dichromates when appropriate radiant energy is absorbed. It may be noted that the dichromates, being orange-red in colour, must absorb strongly in the blue region of the visible spectrum. The action of such energy absorption is to bring about a chemical change in the dichromate which is referred to as *reduction*; this takes place in the presence of the colloidal matter because the latter is capable of aiding the reduction by itself being *oxidized*; such an inter-relationship is often an essential requirement for an effective photosensitive system.

Despite a great deal of work carried out on dichromated systems, there is still no full understanding of the mechanism whereby the

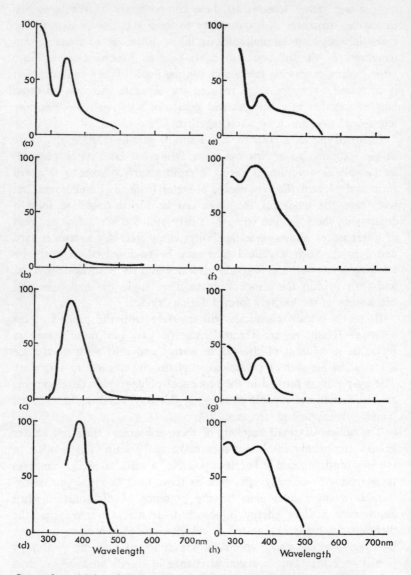

Spectral sensitivity of dichromated systems: *a*, absorption of ammonium dichromate – 1mm thickness of 0.05 per cent solution. *b*, dichromated albumen. *c*, dichromated gelatin. *d*, dichromated gelatin (Autotype G21F paper with potassium dichromate). *e*, dichromated albumen. *f*, dichromated shellac. *g*, dichromated PVA (polyvinyl alcohol). *h*, synthetic vinyl resin with ammonium dichromate.

128

reduction product of the dichromate brings about insolubility of the colloid material. Research continues on this subject, but tends now to concentrate on the synthetic systems such as polyvinyl alcohol/ dichromate mixtures. This lack of knowledge does not prevent successful use of dichromate systems, but does require close adherence to materials and conditions if results are to be predictable.

Disadvantages of dichromate processes

The long-continued and widespread use of the dichromate-based processes is sufficient proof of their practicability and utility. They have two drawbacks: the first is the tendency of chromium compounds to attack the skin and cause 'chrome ulcers'. Common sense, frequent washing and, when necessary, use of rubber gloves should prevent trouble.

The second is the so-called 'dark reaction'. This refers to the tendency of the dichromate/colloid layer to become insoluble *without exposure to light*. Whatever the chemical changes which occur on exposure, similar changes are brought about by heat. At room temperature this process takes place slowly and no problem arises so long as the sensitized material is used while fresh. In commercial operations it is usual to make up sufficient sensitizer for the day's work and discard any surplus. Coated material is preferably used on the day it is coated, but it is possible to hold it over for a few days without detriment. Despite many efforts, no way has been found of eliminating this effect, although the onset of the tanning action has been delayed a few months in some systems. Refrigeration at $-7°C.$ is claimed to remove the risk for an indefinite period.

Applications of dichromate processes

Essentially, therefore, the dichromated-colloid processes are 'do-it-yourself' methods in which the user must carry out the sensitizing of his own material shortly before use. The photomechanical worker would be in the same position as the wet-plate photographer of the mid-19th century, were it not for the pre-sensitized materials which will be described later (see page 167). Even so, the simplicity and reliability of the dichromate type of process gives it many adherents. These are in the fields of:

Photoengraving: line and screened blocks for letterpress printing.
Photolithography: planographic plates, e.g. 'deep-etch'.

Photogravure: etched plates and cylinders.

Silk-screen printing: 'tissue' for preparing the stencil.

Theoretically and no doubt in practice, almost any of the dichromate/ colloid systems could be used in each of these four types of situation. However, trends or fashions have grown up which reflect preferences for one or the other for a given purpose. These preferences are to be respected as the practical wisdom of skilled craftsmen; in photo-engraving it is not uncommon to find process workers whose fathers and grandfathers followed the same skilled occupation. Reprographers will not require to be photoengravers but will find interest in the technique of producing letterpress blocks by metal etching.

Choice of metal for photoengraving

The metals most frequently used for blockmaking are zinc or zinc alloys, and copper. After forming the etching resist, zinc is dissolved away with diluted nitric acid, copper with a solution of ferric chloride. The tendency has been to use zinc for line blocks or coarse half-tone screen blocks, but copper for fine screen work where accurate repro-duction of tonal values is required. There is an interesting reason for this.

To withstand the action of the etching bath, the photo-resist requires to be 'burnt in'. The metal plate carrying the resist image is heated for a few minutes to some 300–350°C. At this temperature the zinc, which commenced in a micro-crystalline form, becomes more coarsely crystalline. This crystal structure of the zinc metal tends to produce uneven rates of etching and therefore the crystal-pattern becomes superimposed on to the image. Polyvinyl alcohol resists are burnt-in at lower temperatures and are therefore preferred for certain purposes to natural colloid layers.

In the case of copper there is little or no change in the structure of the metal during the burning-in step; moreover the action of ferric chloride on copper is more even or uniform so that the etched pattern more precisely follows the resist image. A further reason for the preference in fine screen work for engraving in copper is the better 'etch-factor'. As etching proceeds in *depth*, the etching bath also acts laterally. If this were allowed to continue the image would become undercut and the areas which are intended to print become smaller than they should be. This tendency is less pronounced in etching copper, so precautions against it are less necessary.

Zinc can be etched to an appreciable depth only if the side-walls of each etched area are given protection from the etching solution. Before etching has gone too far, the plate must be washed and dried, then 'rolled up' with an ink, or 'dusted' with a fusible resin powder. Dusting is followed by heating to fuse the powder and bond it to the metal. Dusting and fusing is carried out four times in succession at each stage of etching (once from each direction to give all-round protection to the printing-parts). Etch factor is defined as the ratio between the depth of etch, adjacent to a line, and one half of the loss in width at the top surface of the line. Etching unprotected lines gives a factor of about 2; with protection, factors of 25 or more are obtained on zinc and are considered satisfactory.

So long as copper was available at a reasonable price there were sound reasons for its use but today there is considerable economic attraction in using zinc or zinc alloys; magnesium alloys are also coming into use. Moreover, various improved techniques including the screening methods, and quality of photographic emulsions available have made it possible to achieve on zinc the kind of quality for which copper had previously been indispensable. Perhaps the greatest factor is the technique of etching by machine, known as 'powderless etching', or 'Dow-etch' after the Dow Company of America who pioneered this method. (See also page 173.)

Sensitizing the metal

The sensitizer composition for process engraving mainly uses some form of gelatin as colloid: fish glue, process glue, photoengraver's glue, etc. The engraver tends to learn exactly how to achieve the result he wants with some one type, and continues to use it to the exclusion of all others. The makers of these process glues aim at as consistent and uniform a product as possible, but as with all materials derived from natural sources there is not always the degree of control one would like. Synthetic materials are less open to this criticism, and polyvinyl alcohol is coming into fairly general use for this purpose.

As with the natural colloids, polyvinyl alcohol functions in the senstizer by virtue of its large molecule. It is possible in the manufacture of the polymer, to finish at a given molecular size or at least, to have a preponderance of molecules of one size. This makes the synthetic material more consistent in behaviour and more repeatable in successive batches. Another point in favour of the synthetic material

131

is its freedom from mould or bacterial growth. Gelatin, even before it putrifies, will undergo changes due to mould and bacteria which may influence the behaviour of the photosensitive layer.

The sensitizer for metal need contain no more than the colloid and the dichromate. Additions may be made with the intention of delaying dark reaction or of increasing sensitivity; also resinous material may be incorporated to improve the resist after burning-in. Glue as purchased for this purpose may contain some 40–50 per cent solids; a simple sensitizer might be made from a liquid glue with twice its volume of water and from 1 to 2 per cent of ammonium dichromate added.

Before the metal is coated, it must be carefully prepared by cleaning and polishing, but without scratching. The smooth surface must also be given some treatment which will improve the adhesion of the sensitized film when dry. This treatment may be a slight chemical etching, e.g. a few minutes' immersion in dilute nitric acid. The clean metal is then kept under water until wanted.

Coating may be on a machine or by hand but in either case relies on 'whirling' to give a thin uniform film. A mechanical whirler is necessary for large pieces but can be a nuisance for small pieces. The machine gives the more repeatable results as its speed of rotation is under control and is measured by instrument. In either case, the still wet metal is held face up and sufficient sensitizer poured on to cover the plate; this results in dilution by the water on the plate surface but after pouring off, it can be strengthened-up for re-use. More sensitizer is then poured on and the plate slowly revolved to spread the sensitizer uniformly. In the horizontal machine whirler the plate remains face up; whirling by hand the plate is turned over and rotated face down over a heated plate. After a period of fairly low-speed rotation, the speed is increased to throw surplus sensitizer off the edges and to facilitate drying. Machine whirlers are fitted with fans and heating elements.

The dichromate/colloid sensitizer is heat-sensitive as well as light-sensitive. All being well the plate is dried with moderate warmth which will not induce hardening of the colloid, and is then ready for exposure under the negative. The intensity and duration of exposure are inter-dependent; the minimum exposure is that which completes the light reaction *throughout* the thickness of the layer. It is essential that the resist formed be made insoluble from the top surface of the sensitizer down to the layer in contact with the metal.

Processing and burning-in before etching

While image formation is completed with the operation of exposure, there are several processing stages to carry out before the etching resist is achieved. Where exposed, the gelatin of the glue used has been rendered insoluble; the insoluble parts are those which in the negative transmitted radiation, or were 'transparent' in ordinary language. The photo-image is thus in the pattern which is required to print from the finished plate, and will protect these parts in the etching process. Just as the unexposed plate undergoes dark reaction, the exposed plate continues to react until it has been taken through the 'development' process. In this the exposed plate is washed for a few minutes in water by dish or spray, then immersed in a dye bath which gives the image an intense colour. The washing process may then be followed more easily, if necessary assisted locally by rubbing lightly with a pad of cotton wool. Washing must continue until all non-image parts are completely bare metal. Any residual sensitizer in the non-image parts will prevent the etching process taking place at the correct rate. When washing has been completed the plate may be given a treatment in a hardener to toughen the gelatin image. Burning-in (see page 130) completes the operations prior to the etching proper, and gives the toughened image greater acid resistance.

Cold enamel as engraving resist

Burning-in may be avoided in photoengraving by the use of 'cold enamel'. Shellac or synthetic material replaces the process glue, but the sensitizer must then be prepared in a solvent such as methylated spirit. In coating, less water is left on the cleaned metal and only very gentle heat used. After exposure, unwanted parts of the layer must be dissolved away using the same solvent as in preparing the sensitizer; e.g. in a bath of methylated spirit containing about 0.25 per cent of a suitable dye (methyl violet, etc.). A final rinse in water and if necessary local rubbing over with cotton wool completes the plate for etching.

Advantages of photolithography

The process of preparing blocks for letterpress printing is obviously lengthy, skilled and expensive. The use of photoengraving was first

133

restricted to the provision of illustrations to accompany text, the text being prepared from movable type which had the advantage of being reusable. Today it is becoming common to use engraved plates for text, owing to the advantages of the powderless-etch process.

Photolithography, however, was able to offer quicker and cheaper methods of illustration and its techniques were adaptable to text, paving the way for the modern approach via phototypesetting.

Albumen as engraving resist

The most widely used photolithographic sensitizer for 'surface' plates was dichromated albumen; the term 'surface' implies that the final printing image is in the same plane as the non-printing parts of the plate. This is not strictly true but distinguishes albumen plates from those in which an etching stage, however slight, is relied upon to give a greater differentiation between printing and non-printing areas.

Dichromated albumen has also been extensively used as a photo-engraving resist for line blocks and thus can be considered as a process for more than one application rather than as specifically for the surface plate of lithographic work. In both cases the metal must be clean and coated in a thin uniform layer with the dichromate/albumen mixture; drying should be rapid but at low temperatures, not above 50°C. After exposure under the negative the coated side of the plate is covered with a thin layer of an ink (carbon black and a binder, in turpentine) using a soft pad or a composition roller. Only a grey appearance is obtained and the ink film is not continuous. This fact allows the unexposed parts of the albumen layer to swell and dissolve away on immersion in water.

Gentle rubbing with cotton wool under the water surface encourages the ink to leave the non-image areas and transfer to the image parts; in other words, exposure of the albumen/dichromate layer has produced an oleophilic effect while the unexposed parts remain hydrophilic and water-soluble.

It should be clear from the last paragraph that the term 'hydro-philic' is associated with a 'liking' for or acceptance of water, and 'oleophilic' to similar properties but for oily materials. As these terms are of frequent use in connection with lithography, it would be as well to ensure that they are correctly understood. The British Stan-dard Specification 4277:1968 (Glossary of Terms used in Offset Lithographic Printing) includes the term hydrophilic and its definition as: 'Water attracting. A necessary property of the non-image area

134

of a lithographic plate'. From this starting-point, one easily obtains the opposite meaning as hydrophobic: a dislike for water (as in hydrophobia).

Thus we have a pair of words which can describe the basic situation of lithographic printing; but in addition to describing the position from the standpoint of *wetting* the plate surface, it can be described from the *inking* properties. Then:

hydrophilic = water attracting = oleophobic or oil-repelling
and hydrophobic = water repelling = oleophilic or oil-attracting

While all these terms are found in use, it seems proper to point out that *only* hydrophilic is included in B.S.S.4277 : 1968.

A word of warning is appropriate here. The mere fact that we use terms as above, to describe the lithographic situation, does not mean that by using these words we have an *explanation* of that situation. It is not sufficient simply to have oleophilic image, and hydrophilic non-image, areas. These are merely the basic requirements for a lithographic plate to function. But the printing is done by a correct choice and balance between fount solution and ink. Success depends on the surface tensions between fount and ink, which must act to prevent *spread* of ink away from the image on to the fount solution film held by the non-image areas.

When the albumen process is used for photoengraving, the image obtained must be further strengthened. The warm, dry plate is dusted with resinous powder which adheres to the tacky inked parts but not to the bare metal. The plate is then heated to a temperature at which the resinous material fuses with the ink layer to form the acid resist.

Principles of deep-etching litho plates

Although albumen plates are still used in photolithography, the deep-etch method has largely replaced them on account of greater reliability and longer runs on the press. The method was first described in 1891 but did not attract attention for 25 years or more. By 1930, however, it was widely used and has continued so to date. Of course there is nothing gained in making a plate capable of 100,000 copies if only 10,000 are required; but the reliability of the deep-etch method is such that if the process is available there is always a temptation to use it. An attraction of deep-etch is that although the sensitizer is negative-working like all the dichromated materials, a positive is used for printing-down the image. It is confusing if the deep-etch process is described as 'positive-working'; it is negative-working, but *positive-using*.

Use of anodized aluminium: The metal mainly used for deep-etching today is aluminium which has been anodized – a process carried out by electrolysis but distinct from the milder process of electrolytic graining. For lithographic use the anodized metal is given a further treatment such as boiling in water, which is said to 'seal' the surface. The attraction of metal prepared in this way is that it has a tough hard surface layer, resistant to scratching, and which is also inherently water-accepting. This simplifies operations on the printing-press but creates a problem for the plate-maker, because the water-accepting property has to be removed in those areas which are required to print.

Sensitizing and development: The sensitizer for deep-etch is di-chromated gum arabic, of very simple formula – about 20 per cent gum arabic and 5 per cent ammonium dichromate. Viscosity and density have to be controlled so that consistent coating thickness is obtainable. The coating operation is similar to that for albumen plates, but a greater thickness is required to give protection during the etching stage. As with all materials having a thick coating, there is some tendency in exposure for light spread, which on account of the negative-worker sensitizer, tends to make lines thinner and dots smaller in the final print. Exposure must be sufficient to give adequate hardening of the gum coating.

Although exposure hardens the gum layer, it does not give complete insolubility in water. Development has to be carried out with a liquid of restrained action, the most common being a strong solution of calcium chloride. The time of development depends on the strength of the calcium chloride solution and the working temperature. The quality of the photographic positive from which the plate is made also has a decisive influence on the ease of the development operation, which may require considerable skill to achieve a faithful reproduction of the image on the film positive. The developer is poured on to the plate and spread with a plush pad so that a slight abrasive action is given. A second application of developer is usual.

Etching stage: After development, the plate is etched and numerous formulae for this stage have been published. In the case of the anodized aluminium plate, the effect of the etch is to remove about half of the anodic oxide film. Depth of etch is not more than 0.0003in.; if too great it becomes difficult to ink the plate satisfactorily. The etching solution must therefore be removed as soon as its action is complete, but to wash with water carries some risk of attack on the resist while the etching chemicals are still present. Several washes are therefore given with anhydrous (dry) industrial alcohol, isopropyl alcohol, or other organic liquids.

The etched areas of the plate will now accept the printing ink, but to give a more durable plate the etched image is first lacquered. A lacquer is a true solution of resinous material in a volatile solvent, and dries rapidly by solvent evaporation to leave a continuous layer of the resin. An ink is a suspension of pigment in a varnish-like 'vehicle', and dries slowly to leave a more or less soft layer of material. Thus laquer to be applied to the etched plate should dry to a film with strong adhesion to the metal and of good ink receptivity. After lacquering, a thin coat of non-drying ink is applied. This ink must not harden, and can be dusted with chalk before the resist is finally removed with warm water (about 50°C.).

With the resist removed, the plate comprises a non-printing background where the original anodized surface is unaffected and image or printing areas in which the anodizing has been partially removed, but leaving a good 'key' or bond for the lacquer, which itself provides a good 'take' to the ink. A final treatment with a solution of phosphates, or of phosphoric acid and gum arabic, preserves the hydrophilic condition of the non-printing areas and leaves the plate protected for storage before use on the press.

Other applications of deep-etch principles

It should be clear from this abbreviated description of the deep-etch process for lithographic platemaking that the final result can be considerably influenced by the operator. The development stage especially permits or may demand, a high degree of skilled control to achieve the best result from a given photographic film positive. Numerous proprietary systems of sensitizer, developer and etch are available from manufacturers. These have found application in fields other than lithographic printing, for example cartography and the preparation of metal name-plates.

For example, in preparing proofs of maps, usually in several colours, one can start with a sheet of film, transparent or 'white' according to the requirement. The film is sensitized with deep-etch sensitizer, exposed to the first positive and developed. A dye solution can then be applied and dyes the base film through the image openings of the resist. The resist can then be removed and the whole process repeated for a second, third, etc., colour. The finished article not only 'proves' the accuracy of the various parts of the map, but is itself a very attractive specimen of work.

Gravure and silk-screen methods

The processes of gravure and silk-screen printing also utilize dichromated colloid sensitizers. In these two cases, although it is feasible to sensitize the gravure metal or the silk screen directly, it is more usual to form the image on a sensitized layer carried on a temporary support. Dichromated gelatin is typical and the temporary support may be paper or other sheet material. Because of the dark reaction of dichromated systems, it is usual to purchase the coated sheet as 'tissue' which requires to be sensitized before use. This simply means immersion in dichromate solution for the recommended time, and drying in the dark.

The process of making a deep-etch plate for lithographic printing, or the map-proofing process described, and other variants, achieve 'reversal' images and the technique involved may be referred to as 'reversal processing'. By this is meant that a negative-working photosystem is used to produce a *positive* result from a positive original. Both in the deep-etch plate and the map proof, the image first produced is a negative of the original; the processing succeeds in using this negative image as the means of making the required positive image in lacquer, ink or dye.

Gravure and silk-screen printing have the same characteristic of using a negative-working photosensitizer to achieve a positive result, though the techniques are not 'reversal' in the normally accepted sense. For gravure, printing is achieved by transfer of ink in small recessed cells from the plate to the paper under pressure. Tonal values –i.e. densities intermediate between the maximum and minimum ink weight – are obtained by controlling the depths of the cells and therefore the amount of ink which they each can contain. Thus the problem in making a gravure printing surface is to control the depth to which the cells are etched out of the metal.

Types of gravure printing

For letterpress and lithographic printing surfaces, the plate is made from a film transparency in which tones are achieved by creating a pattern of differently-sized dots; each dot irrespective of size is

138

required to print at maximum 'blackness' or weight of ink in the colour required. Letterpress blocks and lithographic plates are thus made from *screened* films, whereas gravure plates require continuous-tone positives. Although screening is not carried out in the photographic operations, the gravure plate has to be given a structure of minute cells, each one of which can be regulated in its depth of etch. The gravure plate, or tissue if that method is employed, therefore receives two separate exposures. One of these is under a regular screen and secures the walls or partitions of un-etched metal which make the cells. The other is the actual image to be reproduced. The net result is *not* a screened image in the normal sense of letterpress and lithographic work (see pages 120–122).

The conventional gravure method achieves tonal variations by relying on variable depth of etch in cells all of one size. However, gravure technique is not restricted to this classical situation; various methods have been introduced which combine the principle of variable depth of cell with that of variable area of cell. In one such method, the tissue is exposed twice to a continuous-tone positive – once with a contact half-tone screen in position, and again without the screen. In some cases, ink cells of uniform depth are produced and tones rely solely on the cells being of different area. All such processes are of the intaglio, gravure, or recessed type.

Because of the requirement in gravure to print from recessed cells, one cannot obtain a truly continuous printed line; also, text which is to be printed alongside illustrations has to form part of the engraved printing image, and is also affected by the imposition of the cell pattern. On the other hand, the tonal values are especially attractive. These properties of the gravure process give it a distinctive quality, but good results depend very much on the skill used in the etching of the plate. In lithographic printing an economical 'run' is typically 10,000 to 100,000 copies, while 1 million might be normal for more durable plates described as bi-metal and tri-metal. With modern rotary gravure methods, it is hardly economic to undertake preparation of the printing cylinders and setting-up the press, unless more than one million copies are wanted.

Producing the gravure image

The sensitized tissue is exposed under the film positive to produce varying degrees of hardening in the dichromate/gelatin layer. The hardening is directly proportional to the amount of radiation absorbed and therefore to the amounts transmitted by the positive. Where the positive is black there is the least hardening, where transparent, the most. The extent of hardening decides the depth of etch and the weight of ink finally available for transfer. Thus, under the blacks of the positive, the least hardening occurs and the greatest depth of etch, and maximum weight of ink is transferred to the final paper surface.

The hardening process always starts from the *top* surface of the sensitive layer. Where there is to be no etching, the correctly-exposed layer carries areas and lines (the cell-walls) which have been hardened throughout the thickness of the layer. Other parts have various depths of hardening, all starting at the surface and extending downwards through the gelatin layer. The exposed tissue is therefore transferred to the copper surface *before* it is developed, with the top-side of the gelatin layer against the metal, so that all hardened parts of the layer adhere to the metal. The temporary paper support of the tissue can be lifted off by immersing the metal cylinder in warm water. Residual unexposed, and therefore still soluble, gelatin is then carefully washed away, the cylinder dried and prepared for etching.

The etching of copper is carried out with ferric chloride solution, which penetrates the varying thicknesses of hardened gelatin at a rate controlled by the gelatin resist. The etching process tends to have a 'characteristic curve' in a similar way to a silver halide layer, because the *rate* of penetration of ferric chloride is greater in the thin areas than in the thicker areas. In the making of the resist, hardening is directly proportional to the amount of radiant energy absorbed; in etching the metal, double the thickness of resist requires eight times as long time of penetration.

Screen printing stencils

In the silk-screen process, a stencil or mask is attached to fine gauze stretched on a frame; the gauze is laid on the work to be printed and ink on the top side is forced through the openings of the stencil by traversing a rubber blade. Stencils may be cut by hand and attached to the screen, or may be produced photographically from original designs on transparent media or from film transparencies, etc. There are two ways of using a photo-sensitizer: the screen itself may be covered with the sensitive material, or a 'tissue' may be used as in photogravure. The finished design on the sensitized tissue is transferred to the screen in the same way as a hand-cut stencil. Dichromate/colloid systems are used in screen-printing, but other types are applicable.

The advantage of sensitizing the screen direct is that the stencil is completely 'embedded' in and around the openings of the gauze and is therefore very robust. However, screens are re-used repeatedly and a directly-sensitized screen is more troublesome to wash out for re-use. Moreover, as washing-out, re-sensitizing, and exposing must be carried out with the gauze stretched on its frame, these operations tend to be cumbersome. Screens using the tissue method avoid some of these difficulties and are more easily cleaned off for re-use; but the stencil itself is not so robust as that produced on a direct-sensitized screen, nor is detail so accurately retained.

Note again that whichever method is used, the sensitizer is negative-working but is exposed under a positive of the design to be printed.

CHAPTER 9

PROCESSES USING IRON COMPOUNDS

Iron salts form the basis of a number of reproduction processes in much the same way as do the chromium compounds. But where the great value of chromium systems is in the varied use of 'tannable' colloids, the application of iron salts usually involves a specific chemical reaction between the new compound produced by light action, and a second substance, the product of reaction being highly coloured. Compounds of iron are referred to as ferric or ferrous; the two forms frequently being mutually inter-convertible by the processes of oxidation and reduction:

$$\text{ferric compound} \xrightleftharpoons[\text{oxidation}]{\text{reduction}} \text{ferrous compound}$$

The term 'oxidation' initially was used to indicate combination with oxygen but it now has a wider significance and does not necessarily involve reaction with oxygen itself.

One group of iron-salt processes depends on the highly coloured compound formed between ferrous salts and the so-called ferri-cyanides, or ferric salts and the ferrocyanides. Thus ferrous-ferri-cyanide is a deep blue compound also known as Turnbull's Blue. A rather darker blue compound is the Prussian Blue of the paint-box, which is ferric ferrocyanide. If a material is prepared with a light-sensitive ferric salt, it can in theory be used to give either a negative or a positive copy. In the former case, ferrous salt produced by exposure would be used to react with a ferricyanide; in the latter case, the residual ferric salt (unaffected by exposure) would be reacted with a ferrocyanide.

Herschel's blueprint process

The blueprint process, dating from 1842 and widely operated up to quite recent times, was the negative-working version. To stabilize

142

the print it had to be washed in water and the base-paper selected took this into account. 'Ferro' paper is a strong white paper (i.e. untinted), chemically pure, sized and calendered to give the right degree of penetration for the sensitizing solution. Unlike silver halide or chromium-salt systems there is no colloid substance used which will dry to a film on or in the top surface of the paper sheet. The base paper for blueprint process must therefore restrict the penetration of the sensitizer into the lower parts of the sheet; if the solution penetrates too deeply, part of the sensitizer salts will be hidden beneath the paper fibres and not therefore receive the energy by exposure which brings about the required change of chemical state.

The base paper can be prepared with pre-coats which are designed to control penetration, to avoid revealing the structure pattern of the paper in the final print, and also to increase visual contrast of the print by concentrating the blue pigment of the image in a thinner layer.

The sensitized paper merely carries a ferric salt; the blue pigment is produced by application of a solution containing, say, potassium ferricyanide. The photoreduction of the ferric salt takes place only in the presence of an oxidizable substance, which can be the cellulose fibres of the paper or the starch, gelatin or other sizing used in paper manufacture. Better still, the ferric compound may be a salt of an organic acid which is itself capable of oxidation. Herschel used ferric-ammonium citrate but similar tartrates and oxalates are also suitable. Herschel also added to the sensitizing solution the ferricyanide required to produce the blue pigment from the ferrous compound produced by exposure. Thus the sensitized and dried paper contained all the necessary chemicals. In this case the effect of exposure is to form a pale-coloured image, as the colour-reactions only take place in the presence of sufficient water. When the exposed paper is immersed in water, the blue image reaches full strength and un-reacted chemicals are dissolved away.

The blueprint process is a very simple one to operate and larger users frequently sensitized their own paper on comparatively uncomplicated machines – a roller which applied sensitizing solution, a glass bar to scrape off the excess, and a heated chamber to dry the paper. Equipment for making the prints included automatic washing and drying, so that the copies were flat and ready to handle. Against the simplicity and reliability of the blueprint are the negative image – white lines on a blue background – and wetting of the print, with consequent expansion, and shrinkage on drying. In the USA especially

143

the process has remained in use mainly because of the availability of advanced equipment which automated the washing and drying operations. Alternative names for blueprint are ferro-prussiate (an obsolete name for the ferrocyanides) and cyanotype.

Positive blueprint processes

To achieve a positive print it is necessary to sensitize with the ferric salt, expose, and 'develop' with a solution of potassium ferrocyanide. The first practical working method was that of Pellet (1877); others came from Collache (1880) and Pizzighelli (1881). These varied only in the method of preparing the sensitizer, which included gum arabic.

Ferazo process

In the 1930's a method of producing negative blueprints without washing appeared under the name Ferazo (H. D. Murray and D. A. Spencer). This achieved yellowish white line images in a full-strength blue background by applying a nickel salt solution to the print surface after exposure. The effect is to render the ferricyanide non-reactive with the iron salts remaining after exposure. The solution was applied mechanically from a roller rotating in a trough of the liquid, in much the same way as semi-wet diazo prints are processed. In its final form, this process gave prints with brilliant blues and clear white lines.

'Brownprint' processes

When a ferrous compound produced from a ferric compound by actinic light reverts to the ferric form by an oxidation process, it can reduce platinum, gold and silver salts to the metal. Several processes on this basis have been devised, of which only the Van Dyke remains of interest today.

The paper for the Van Dyke process carries silver nitrate and a ferric salt, with possibly other substances to accelerate or 'sensitize' the reactions. The image is formed by the ferric⟶ferrous change, and the silver, etc., salt is used to develop the image in a more permanent and visible form – any light-sensitivity it has is of little or no significance. The brownprint is more common in the USA than in Britain and is mainly of value as a negative from which blueprints can be made. It is also said to have advantages as a means of proofing for multi-colour lithography when working from negatives.

Iron salts with colloid materials

Ferric salts will harden gelatin, so that on exposure to light and conversion to ferrous salts, the gelatin may revert to a soluble form and give the basis of a wash-away system. The resultant image would be a positive of the original, whereas the use of dichromated colloids results in a negative image. There does not appear to be any widespread interest today in the use of iron salts in this way.

A rather different type of system is typified by the TTS or 'true-to-scale' process. This name comes from the fact that the copies are made without the paper or other base material being wetted and dried, therefore not changing its shape or size as a blueprint might. The TTS process as a whole is a lithographic operation, though the imaging function is performed by iron salts.

The TTS 'table' is a slab of gelatin composition, prepared hot and cast on to a large flat surface. The composition contains ferrous sulphate. The first stage is to expose a sheet of conventional blueprint paper to the original drawing, then without wetting, place it face down on the composition. In the areas of the blueprint which were protected from light by the original lines, the unreacted ferricyanide remains unchanged. It diffuses out of the paper into the gelatin layer and reacts with the ferrous sulphate of the composition. The reaction brings about a change in the surface characteristics of the composi-

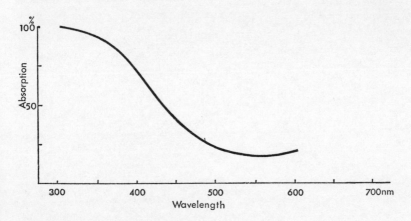

Spectral sensitivity of blueprint paper.
Suzuki, Matsumoto, Harada and Tsubura: *Photographic Science and Engineering* **12**, 2–16 (1968) (reproduced with permission of the Society of Photographic Scientists and Engineers).

tion, so that in these areas (which correspond to the lines of the original) the gelatin is tanned and accepts greasy material such as printing ink. After a sufficient time for the reaction to take place, the blueprint is peeled off and a lithographic ink applied by roller to the composition. Provided that the blueprint received sufficient exposure to convert all the ferricyanide to Turnbull's Blue in non-image areas, ink will adhere only to those parts of the composition which correspond to the lines. Then by pressing a sheet of paper or cloth on to the composition, ink is transferred and gives a positive copy of the original.

The TTS process had a great vogue in the first thirty or so years of the present century, ousting Pellet's positive process (see page 144). Attractions are the pigment image on untreated base – frequently tracing cloth, bristol board or fine-quality paper. The copy is thus unaffected by chemical changes which might occur in a sensitized material if imperfectly processed. It enjoyed the name Ordoverax, which name still appears in American texts.

By successive inking and transfer a series of prints can be made, this amounting to a *direct* lithographic process. The composition may be melted and re-cast for further use, provided water loss is made up. Preferably some new composition is added at the same time. Velography is a basically similar process but uses a paper coated with potassium dichromate. A number of writers regard Ordoverax and Velography as synonymous, but this is incorrect.

CHAPTER 10

PROCESSES USING DIAZO COMPOUNDS

The groups of processes hitherto described depend upon compounds of metals: silver, chromium, iron. Their resultant images are metallic silver, pigmented colloid, pigmented printing ink or the blue pigments from iron. The so-called diazo compounds have given rise to an equally diverse range of photoreproduction methods but without making use of light-sensitive metal compounds. They have their origin in the dyestuff industry and their end-point is frequently a dyestuff image. In more recent years they have also been used in photo-mechanical situations in which case the end-point again is the pigmented printing-ink of the final copy.

Nature of diazo compounds

Diazo compounds as used in photoreproduction are derived from benzene, the basic chemical of the coal-tar dyestuff industry. Benzene itself need not be obtained from coal-tar and is only a starting-point for the complex substances actually used. The chemist's shorthand for the compound benzene is a hexagon: ⬡ which merely indicates a ring of six carbon atoms, to each of which is attached a single hydrogen atom. More complex structures are made by substituting other atoms, or chains of atoms, for one or more of the six hydrogen atoms of benzene. The significance of the double lines in the hexagon, is that each of the carbon atoms in the ring has available a 'spare' or unused linkage. Each pair of adjacent carbon atoms therefore forms an additional or 'double' bond, but the ring structure gives benzene a greater stability than the straight-chain compounds with double bonds which figure in photopolymer systems (see page 182).

In a diazo compound, one substituting group contains two nitrogen atoms, hence diazo (from the French *azote*, meaning nitrogen). This diazo grouping is chemically very reactive and gives a route to the

147

laboratory preparation of many other types of substance. It is also sensitive to both light and heat. These properties made the first diazo photoreproduction process possible, while heat sensitivity imposes certain restrictions.

Diazo compounds were first discovered by Peter Griess in 1858, and the dye coupling reaction was first used by him in 1864. His work gave rise to tremendous developments in the manufacture of synthetic dyestuffs. It was not until 1881 that Berthelot and Vieille reported that diazo compounds were light-sensitive. In 1885 (Dr. West) and 1889 (A. Feer) described possible photographic processes using diazo compounds; but diazotype reproduction as known today commenced in 1890 with the so-called Primulin process of Green, Cross and Bevan. The Primulin process gave a positive image but though workable did not become a commercial method. It remained neglected for nearly thirty years, though light-sensitivity of the diazo compounds continued to attract interest. Eventually came the patent of Kögel in 1920. Much further research and development were pioneered by the Kalle Company of Wiesbaden, Germany. Important contributions also came from the Van der Grinten laboratories in Holland (1927 onwards).

As with the systems already described, the diazo compound alone does not constitute a functioning photoreproduction method. It provides the means whereby an image can be created; but the image then requires to be made permanent, and to be plainly visible. One reaction which makes this possible is known as dye-coupling, and the diazo compound may be considered as a half-way stage towards the formation of a dyestuff.

Principles of dye-coupling

The substance with which a diazo compound couples to give a dyestuff is usually referred to as a coupling component. Coupling components may be of various types, and different combinations of a few diazo compounds and a few coupling components give a diversity of possible final dyestuffs. It is this which has given the diazo processes great flexibility and has made it possible to devise reproduction materials to fit a wide variety of differing requirements. For the same reason, there is a danger of using a diazo material for some purpose other than that for which it was intended.

When a diazo compound is either heated or exposed to actinic light it undergoes chemical change and loses the ability to form dyes by the coupling reaction. It is usual to speak of the diazo compound as having been destroyed or decomposed, by which is meant that the diazo compound has ceased to exist as such. For the purpose of a photoreproduction system relying on dye formation, it is sufficient to recognize that the diazo compound no longer exists where it has absorbed sufficient radiation energy; but what in fact has happened is that the substituent group in the benzene ring containing two nitrogen atoms – the 'di-azo' group – has been split off from the ring. From the chemist's point of view the reactive diazo group has ceased to exist but the substance is not destroyed in the sense that the benzene ring itself is disrupted or broken down into simpler compounds.

On the contrary, the benzene ring survives the radiation effect and reappears in a new substance (see below). While these changes occur as a result of the absorption of energy into the diazo compound molecule, a second substance is required to enter into the situation and to promote the formation of the new compound. This second substance is simply water. 'Dry' paper can contain up to 7 per cent of water, which explains why dried diazo papers which have been exposed to atmosphere and therefore allowed to regain moisture show an increased 'speed' or sensitivity to light. These effects are usually noticed on the edges of sheets and rolls or the top few sheets in a stack; so that a careful operator takes steps to avoid such effects if of importance to the work he has in hand.

Coupling components come from a number of chemical groups. One of the largest is the phenols, which again consist of a benzene ring carrying one or more substitutions; to be a phenol, at least one substituent must be a hydroxyl group, shown as – OH. The coupling component, or 'coupler', also contributes coupling energy to the process of dye formation, so that more than one factor arises in the choice of a particular diazo-coupler combination.

The reactions and changes so far discussed are conveniently summarized diagrammatically:

Action of actinic light

Diazo compound   Phenolic compound

$$\langle\bigcirc\rangle\!-\!N\!:\!N\!\cdot\!X + H\!\cdot\!OH \longrightarrow \langle\bigcirc\rangle\!-\!OH + N_2 + H\!\cdot\!X$$

Water

N = nitrogen. X = a terminal acid. Apart from the conversion of the diazo compound to a phenolic substance, nitrogen (gas) is liberated.

Coupling reaction

Dye coupling is assisted if the HX acid is absorbed by having alkali present.

The changes represented above form the basis for three different methods of operating a process with diazo compounds:

1. The residual diazo compound after exposure may be coupled to give highly-coloured azo dyestuff. This would be a positive or bleach-out process. See diagram p. 151
2. The phenolic compound produced by exposure could be used as the coupling component for dye formation, giving a negative process
3. The nitrogen formed on exposure might be used to form the image, again giving a negative process.

All these possibilities have been used with considerable ingenuity; the first method is that most commonly met and will be dealt with more thoroughly.

Primulin process

The Primulin process (*Green, Cross and Bevan 1890*) was of the first type mentioned above and showed a similarity in approach to the earlier photoreproduction processes. The inventors made the diazo compound on the actual material to be used for the copy, reminding us of the precipitation of silver halide in the wet collodion plate process. The most general method of making diazo compounds is to act upon an amine with nitrous acid. In diagram form, starting with aniline:

aniline
(or: amino-benzene)

benzene diazonium
chloride

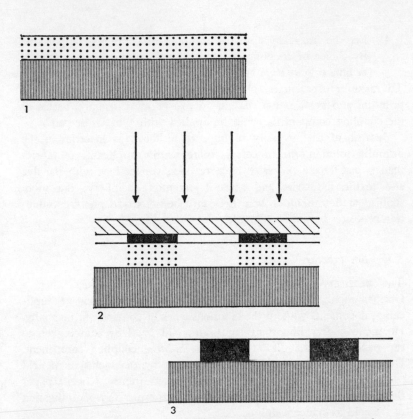

Positive image formation by dye-coupling in diazo material. 1, unexposed material (sensitized film). 2, exposure under original with diazo remaining only where protected. 3, diazo image "fixed" by dye-coupling.

As the diazonium chloride is heat-sensitive, this reaction is carried out in the cold (less than 5°C.) and the solution must be kept cold to avoid decomposition. The diazonium chloride is shown on the assumption that nitrous acid has been liberated from sodium nitrite and hydrochloric acid.

Primulin is a more complex substance than aniline, but it contains an amino group which can be diazotized. The Green, Cross and Bevan method was therefore to:

1. Soak the base material in a solution of primulin, drain, rinse and dry.
2. Immerse the material in a solution of nitrous acid, rinse and dry (in the dark).
3. Expose under a pattern or design.

4. Fix the image by applying solutions of coupling components. By choice of coupling component, red, yellow, orange, purple, or blue colours were obtained.

To make transparencies, glass could be coated with a solution of primulin and gelatin. For designs in two or more colours, pastes of the coupling components could be applied with a brush or pad.

Because of the necessity to impregnate the base material in the primulin solution, the process is more readily applicable to fabrics than paper. Diazo processes have retained connections with the dye and textile industries and share a common chemistry; but more sophisticated techniques were necessary before diazo photoreproduction processes became acceptable.

Kögel dry process

The necessary further evolutionary stage was achieved when a German monk, Brother Gustav Kögel, required a method of duplicating documents without the laborious work of copying out manually. He made a very important contribution in 1920 by bringing on to the paper both the diazo compound *and* the coupling component. Dye formation by coupling was prevented by the inclusion of acid (dye coupling is encouraged by alkaline conditions). Kögel's paper therefore required only the two steps of exposure and neutralization of the acid to induce coupling.

Neutralization was brought about by placing the exposed paper in a box containing a dish of ammonia solution, so that the paper surface was exposed to the ammoniacal atmosphere. The print at no time required to be wetted and emerged from the ammonia box ready for use. Such a 'dry' process was a significant advance in practical convenience over any other process known at the time.

Two-component systems

Systems using materials which carry both the diazo compound and the coupling component, are known as two-component systems.

Premature dye-coupling can be prevented by the use of acid in the paper or different forms of diazo compound can avoid this. Nevertheless, coupling takes place sooner or later, especially if the paper is exposed to atmospheric moisture. When it leaves the factory, diazo paper has been dried to a moisture content of 4.5 per cent or less; according to grade, shelf life in the unopened condition may extend up to two years in a cool dry atmosphere and average European conditions.

As diazo compounds are heat-sensitive, for practical processes they need to be stabilized against heat decomposition. This can be achieved by forming double salts of the diazonium compound, frequently that with zinc chloride. Other metals can be used, and certain organic compounds. The stabilized compounds may be dissolved in warm water, and the solutions kept warm ($25-30^{\circ}$C.) in order to avoid crystallization of low-solubility components. Nevertheless, the sensitized paper remains heat-sensitive to some extent, and it is very desirable to hold all stocks of material under cool conditions. Refrigeration is an unnecessary refinement, though occasionally used when a diazo-sensitized product is required to remain in perfect condition over a long period – say for purposes of long-term comparisons, or to hold a speed-standard in manufacture.

With two-component papers, the colour obtainable is decided by the coupler(s) incorporated. Blue, red, brown and black lines are usually obtainable, and green is used to a very small extent. Black is obtained from a mixture of coupling components because no single azo dye is known which is acceptable as black. Thus dark blues or purples are shaded by incorporating one or more couplers giving yellowish dyes. This raises the complication that when the exposed paper is brought into the ammoniacal atmosphere the *rate* of the coupling reactions will probably differ for the different components. The reaction rate is influenced by moisture and temperature as well as by the concentration of the ammonia. A black-line diazo paper must be designed to give a well-balanced black under a variety of coupling conditions.

Ammonia processing by machine

The variables in ammonia processing can be controlled to some extent by using a machine. We speak of 'developing machines' and 'development', but we should distinguish between the process of dye-coupling involved in a diazo process, with that of chemical development of a latent image in a silver-halide process. In the latter case, the choice of developing solution, and of conditions (time and temperature) can influence very considerably the nature of the final silver image. In the diazo case, the work of image formation is carried out during exposure and all residual diazo compound is utilized in dye formation. This process is 'critical' in the sense that it must be correctly carried to completion, but not in the photographic sense that development is used to influence the nature of the final image.

Diazo compounds used in photoreproduction are yellow in colour, i.e. they absorb visible blue light just as the dichromates and ferric salts do; after exposure the residual products are colourless. The diazo image is therefore visible after exposure but only as a yellow line. By dye-coupling the image becomes insensitive to light and is made much more easily visible. Dyes do fade on prolonged exposure to light but this is 'sensitivity' of a very different order to that of the diazo compounds used in these processes.

Despite a great deal of engineering ingenuity, the performance of ammonia machines may be somewhat short of the degree of perfection desirable to produce blacks with absolute consistency. This springs from the fact that for convenience, the ammonia is purchased and supplied to the machines as a solution of ammonia in water ('ammonium hydroxide'). It is the function of the machine to extract from this solution the maximum amount of ammonia gas while rejecting the majority of the water. The usual arrangement is to supply the solution at a metered rate to a heated chamber, and to regulate the supply of heat to avoid evaporation of more water than is thought necessary for a particular purpose. Modern machines achieve a high degree of control and repeatability in this respect.

In the case of blue-, red-, or brown-line papers there is no great difficulty because they contain usually only a single coupling component. But with black-line papers, different shades of black may be seen in prints made when the machine is warming-up as opposed to when it has been running several hours. Experience is required to plan the day's work in the print-room to best advantage.

It is quite possible to supply the machine with gaseous ammonia from a cylinder, and the small amount of water required for quick reaction separately. Although this method offers the convenience of avoiding the handling of ammonia solution in the print-room, it does not yet appear to have resolved the problem of absolute uniformity of conditions. In the last resort, the most nearly perfect results still depend on the print-room operator's skill and knowledge of his product and equipment.

Contrast of diazo process

The ordinary base papers of the diazo process are high-quality white or tinted paper, sometimes surface-sized on the paper-making machine in order to restrict the penetration of the diazo-sensitizing solution. The sensitizer is usually formulated to give a high-contrast reproduction, especially as so much modern drafting is done in pencil. By suitable preparation of the base paper, the natural contrast of the sensitizer may be reduced or increased. In the first case the paper is given a thin surface coating of a white pigment, and the sensitizer distributes itself among these particles to give a 'softened' effect. In the second case, the paper may be given a clear lacquer surface which has the opposite result: the sensitizer is then restricted to a thin layer in the top surface of the lacquer and is isolated from the paper base. The diazo process cannot be 'controlled' in the sense that a normal (silver) photographic process can by manipulation of the development process.

Two-colour diazo paper

As the diazo processes operate by the formation of dyestuffs, it is natural to think of colour reproduction. This is not difficult to achieve but has no importance now that modern colour photography processes have advanced to their present level of quality and convenience. Some interest attaches to two-colour diazo papers, in which the intention is to produce a copy with certain items picked out in a contrasting colour to the main parts of the print. Such papers must contain two separate diazo systems and the original(s) must have some means of affecting the two systems independently; for example, writing or typing can be in two inks or with two ribbons.

The most successful two-colour paper requires two separate exposures under each of two originals but with a red or yellow filter superimposed as the means of separating the exposure of the two diazo compounds.

One-component diazo process

One-component materials follow the pattern of the Primulin process in that the diazo compound is alone on the copying surface and that fixation and development are achieved by applying, after exposure, a solution of the coupling component(s). The Dutch firm Van der Grinten introduced successful one-component materials in 1927. The choice between the two types is partly that of personal preference, although in some areas the dry process is discouraged by legislation against the venting of ammonia machines to atmosphere (as in the light wells of large buildings). So against the objection to ammonia must be balanced the inconvenience of making and using the solutions for one-component products, and the necessity of frequent cleaning of the machine.

The dye-coupling in two-component systems is brought about by neutralization of the acids included to preserve the material before use. An advantage of ammonia for this purpose is that any surplus ammonia gas (which brings the paper to an alkaline condition) can evaporate off and bring about a *print* which is neutral. This is important where prints are to be exposed to much daylight, because the print background yellows more readily when in alkaline condition.

The early one-component papers suffered in that the coupling component had to be applied in an alkaline solution; the finished print retained any surplus alkali and therefore tended to become yellow or brown in daylight. The alkali requirement can be reduced by using diazo compounds of higher coupling reactivity and compounds were found which would couple in neutral conditions or even when slightly on the acid side of neutrality. Hence one-component papers are of two types, so-called 'alkaline development' or 'neutral/acid development'. The latter not only discolour less, but may even improve their background whiteness with exposure to daylight.

With all one-component diazo materials, it is important that the dye-coupling should take place as quickly as possible. This is because the diazo compound which forms the image after exposure is usually water-soluble; therefore when the solution of coupling component is applied the diazo compound tends to re-dissolve and to move away from its original position. Only if coupling is immediate, or the diazo compound insoluble in water, can this be prevented. One-component systems are therefore frequently judged by the amount of 'line-bleed' which occurs. A paper which shows very little line-bleed when properly handled with the minimum necessary solution applied, might well show objectionable bleed if wetted with too much solution. Hence machines for applying the solution are designed to apply a metered amount; care must be taken not to fall below this amount or the final dye-density of the image will be reduced.

In contrast with print-processes requiring a washing stage, one-component materials were first referred to as 'semi-wet'; but with greater attention to the quantity of applied solution, 'semi-dry' or 'moist' has become the more acceptable designation. As with machines for processing the blueprint, modern semi-dry equipment includes a heating arrangement so that the prints emerge dry and ready to

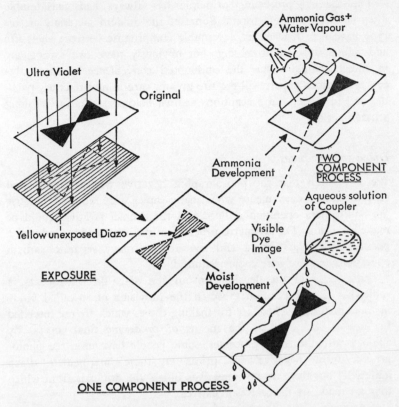

Ultra Violet

Original

Yellow unexposed Diazo

EXPOSURE

Ammonia
Development

Moist
Development

Ammonia Gas +
Water Vapour

**TWO
COMPONENT
PROCESS**

Aqueous solution
of Coupler

Visible
Dye
Image

ONE COMPONENT PROCESS

Comparison between two-component and one-component diazo-sensitized
products: dye-coupling by "dry" method and by "moist" process.

handle. Too rapid drying can slow down the dye-coupling reaction,
or stop it before completion.

Thermal diazo method

There is one other alternative to ammonia processing – the so-called
thermal diazo system, which uses two-component papers containing
the usual inhibiting acid but also containing a substance which *when
heated* will break down to liberate alkali and thus induce dye-coupling
to take place. There are obvious attractions in a material which in this
way can carry not only the chemicals of the photoreproduction
system but also all reagents required for the processing of the exposed
print.

The thermal processing principle has always had considerable attention from manufacturers. Some of the modern materials of this type have now reached an acceptable compromise between shelf life and temperature of processing, but obviously these two aspects are in direct conflict. Thus the undoubted convenience of thermally-processed diazo papers will require greater care in manufacture, packing and in usage, and precautions against holding stocks for too long a usage period.

Diazo intermediates

We have seen that the photographic negative was welcomed and recognized as a convenience when many copies were required, because the one camera operation sufficed and the contact positives could be made at leisure. The Daguerre process, on the other hand, produced a positive from the camera and required as many separate camera operations as there were copies required.

Quite early in the development of the diazo-based products, a somewhat similar position arose in the provision of so-called *intermediate* materials. These are for making copies which are not intended for use as such, but as the means of producing final copies. By analogy with the silver processes, some people have used the photographic term 'negative' for prints on these intermediate diazo materials; but they are not negatives unless the originals from which they are made are themselves negatives.

In principle the diazo intermediates are no different from the two- and one-component products already described. They use a base material of paper, cloth or film, the first two being given a variety of treatments to give a greater degree of transparency to actinic radiation. The image colour is usually sepia, which gives the greatest actinic density to the image when used to make further diazo reproductions. There is a demand for intermediates with a black line, more especially where good readability is necessary in addition to reproduceability, or where the intermediates may be microfilmed, in which case it may be that the black image will be better 'picked up' by the camera and film emulsion.

The main functions of an intermediate are to serve as a step in the total draughting process, or to supply 'masters' of significant drawings to associated or subsidiary companies who are then in a position to make numerous final drawings without these being sent through the post, etc. The availability of diazo intermediates has had a marked influence on drawing-office and reprographic techniques. For example, the technique of exposing an intermediate from a drawing requiring revision, then giving a second exposure under a mask before development, completely removes unwanted parts without erasure or chemical correction.

The earliest intermediate materials made use of a thin (40–45 gsm) well-beaten base paper under the name transparent, and of the tracing-cloth of the drawing-office. While these provided adequate duplicate copies of an original drawing, the base structure tended to re-appear in the final print. Acetate foil was also sensitized for this purpose but was not an ideal material and of course a good deal more expensive. Simple transparent paper could be improved if the stock used had a high proportion of rag (cotton fibre) and "all rag" papers also came into use. Both types could be further improved in transparency by impregnation with suitable resins.

Lacquered materials

More progress came with lacquered papers and cloths. By carrying the sensitizer in the lacquer layer, it became easier to erase the dye image. Thus if a complex drawing required a certain amount of amendment, it could be copied on one of these materials, the unwanted parts erased and the new work added – with considerable economy in draughting time.

Increasing interest in such techniques led to the demand for adequate drawing surface(s) on the intermediate, usually on the unsensitized side, but sometimes on both. This is because the best use of an intermediate comes from always printing in face-to-face contact; thus the drawing is laid face down on the intermediate material, ensuring the best possible reproduction of the line at low background density. When making the next contact print, the intermediate must be turned over to be image-side down in order to bring about lateral reversal, and so retaining the best conditions for exact line reproduction. Thus new work must be drawn on what is then the top side of the intermediate.

Intermediates may well be made one from another in successive stages of a project and another essential requirement is that the lines of the drawing should not only be retained, but also be free from confusion with the background pattern. 'Line spread' and 'infilling' is encouraged unless background pattern is smooth and its density kept low.

159

Polyester materials

Technology has kept pace with the increasing demand for high-quality intermediate materials, principally in the adaptation of polyester film, which is very resistant to chemical attack and enjoys high mechanical strength. Thus whereas cellulose acetate film had to be 0.005in. thick, polyester film of 0.003in. thickness rapidly found acceptance while 0.002in. and 0.0015in. have sprung into more recent prominence. Polyester is a somewhat expensive raw material, but in

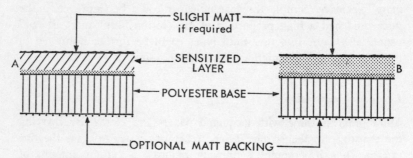

Comparison between diazo-sensitized (A) and silver-sensitized (B) film products.

these lighter gauges the overall cost of the finished article is of the same order as more conventional products such as paper and cloth to which preparatory treatments have had to be given. The polyester product, however, has considerable advantages in use and performance.

Improvement in polyester film-based diazo intermediates has raised the practical question of whether such materials could be used instead of silver-sensitized products. Apart from the attraction of lower price, there is no necessity for dark-room processing and no washing and drying periods. The crucial question is whether the diazo product can give an adequate density in its image; and the answer is predominantly bound up with the *purpose* to which the intermediate is to be put.

For example, in photoengraving and photolithography, it is usual to print-down onto sensitized metal; the sensitizer used on the metal is frequently dichromate-based and negative-working. The silver negatives made for this purpose have as high a density as possible to give the greatest safety-margin in printing-down. A diazo image in the same situation would not have the same margin of safety against over-exposure; but used with understanding it can be adequate for the purpose. If the metal sensitizer itself uses a diazo compound, then there might well be a comfortable safety margin; while the choice of light source for printing-down might increase the margin by 100 per cent or more.

Required properties of diazo compounds

Comparatively few of the large group of diazo compounds known are of use in practical photoreproduction processes. Those proven to be useful must show a combination of all or most of the following properties:

1. High sensitivity to available radiation sources.
2. Sufficient stability in manufacture and use.
3. Ability to produce sufficiently light-stable dyes of suitable colour.
4. Sufficient coupling activity.
5. Minimum background discoloration in the print.

To these must be added that of easy availability, or preparation from available starting materials, because one of the attractions of normal diazo-sensitized products is their relatively low price.

Standard and fast diazo compounds

Diazo compounds chosen for normal print-papers are three or four closely-related members of one group. These are referred to as 'standard' diazo compounds to distinguish them from other compounds coming into use in fairly recent years and known by 'fast', 'super-fast' or similar descriptions. It is important to understand in what respect these two types differ, and to realise that the faster types require more careful handling.

The standard compounds are yellow, not deeply coloured, and have a maximum sensitivity in the ultra-violet region at about 370–380nm. This allows the papers to be handled in normal room lighting without much danger of spoilage, i.e. the handling *time* is not long enough for the diazo compound to undergo significant amounts of photo-bleaching. The 'super-fast' compounds have their peak absorption (maximum sensitivity) at around 400nm and are more definitely yellow in colour, even approaching orange. This means that more of the visible region is being absorbed. The net result is that delay in

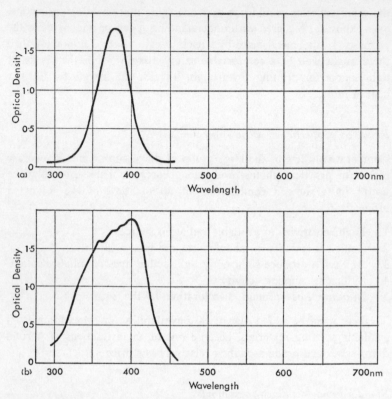

Spectral sensitivity of diazo compounds of general use in
copy-papers: *a*, typical of use in two-component systems. *b*, typical of use in
one-component systems. Based on *Photosensitive Diazo Compounds*
by M. S. Dinaburg (Focal Press, 1967).

handling, unduly bright room lighting, or stray light from ultra-violet
sources, may lead to photobleaching and therefore a reduction in the
ultimate image density. Such materials should always be kept in a
closed box or a drawer, and only the immediate working quantity
should be taken out and kept face down until use. The speed rating
of the paper is not in itself a sufficient guide in this respect because
two papers might carry the same speed rating, the one being of
'super-fast' type and therefore giving a higher density of line than
the other. It would also have the higher sensitivity to visible light
and therefore require the precautions mentioned.

In the search for higher image density intermediates, the super-fast
diazo compounds have been invaluable. The intermediate has to
provide a *functional* image rather than a purely visual one, making

precautions even more necessary, especially because light can penetrate sideways into a stack of film pieces even if the top of the stack is adequately covered.

Other diazo film products

While the major interest in polyester film products is in sepia or black images with draughting surface on at least one side, there is a demand for film products which are completely clear. They may be used as overlays for multi-colour display, or for demonstration purposes on overhead projectors. An increasing interest comes from the proving of colour separation positives in colour-printing, before the printing-plates are made so that it is possible to check, in a matter of minutes, whether the separations are capable of giving a satisfactory result. Use of polyester film is advantageous because there should not be any risk of the various images being out of register through expansion or contraction of the base material. The stability of polyester film makes it fully acceptable for other critical work such as the proving of the separate positives in map-printing, to ensure 'fit' or correct registration.

Negative-working diazo systems

As briefly mentioned above, it is possible to use the action of actinic light on a diazo compound in several ways. For example, some diazo compounds will dye-couple with the phenolic substance produced by their own decomposition; thus dye-coupling can only occur in the areas where radiation has been absorbed.

Other negative-working systems are based on so-called diazo-sulphonates, which are soluble in water and 'passive', i.e. temporarily lacking the ability to couple. The passive compound is converted to its active form by exposure to light. This principle can be used for negative-working diazo materials, which are convenient in that dye-coupling can proceed as exposure takes place; but for permanence the print must be fixed by washing away the residual unchanged diazo-sulphonate – otherwise the print would darken all over under light action and the image would be lost in the overall 'fog'.

Vesicular diazo method

Another negative-working system devised by the Kalvar Corporation in 1957 uses the nitrogen gas given off when the diazo-compound is

163

acted upon by light. The method is attractive because no chemical reactions are involved and after-treatment does not require gases or solutions. It is used for producing film copies (transparencies) from negatives. The transparencies are intended to be viewed by projection because the image is an 'optical' one and requires the specific illumination conditions of a projector to give it full visibility. Similar products are being evolved for other purposes, but also require illumination with parallel-light beams if they are to function as intended.

These 'vesicular' processes rely on a thermoplastic resin film to carry the diazo compound, which must be dispersed within the film and not restricted to the upper surface as is desirable in the normal diazo intermediate relying on dye-coupling. The action of actinic light generates nitrogen gas which is trapped within the resin film; but on warming the film to the softening-point of the resin, the trapped gas expands and creates a multitude of small spherical bubbles within the resin layer. The film re-hardens on cooling and thus automatically 'fixes' the bubble-image. The resin still contains some diazo compound which has not been acted upon by the exposure and which generates nitrogen as and when it receives suitable radiation. But this takes place slowly and fails to form an image as the resin layer at room temperature is not sufficiently plastic, nor is the nitrogen gas under pressure, i.e. requiring to expand. Eventually, the nitrogen 'leaks away' and the copy is fully permanent though still susceptible to unintentional damage by heat.

The vesicular image contains no pigment or other colouring matter, being visible through what is known as 'internal reflection' at the surface of each bubble. The effect of internal reflection may be seen if a glass tube is immersed in water. If water enters the tube, the glass is virtually invisible at those parts; but where air is trapped inside the tube it appears to be silvered as a mirror. Because the effect relies on a glancing angle of incident light, the bubble image must be viewed in a beam of light if maximum density and contrast are to be obtained. The diagram on page 166 shows how a vesicular image of only 0.5 density under diffused light, is effectively 2.0 or more under projection conditions. Moreover, the effective density is greater as the aperture of the projection lens is decreased.

Recently published work shows that image visibility relies on the crystalline structure of the resin at the surface of the bubble; also it is said to be possible to increase the effectiveness of the image by inclusion of salts which might well influence the condition at the bubble surface. Some additions may also produce an opacification

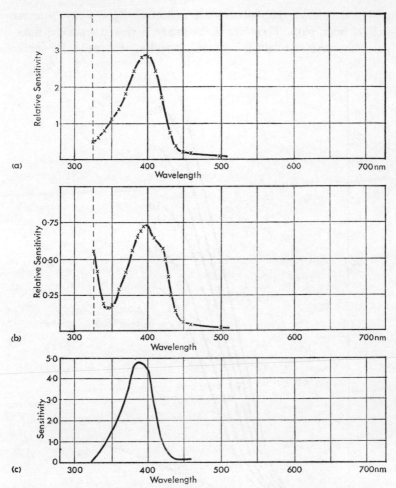

Spectral sensitivity of some specific diazo products: *a*, film intermediate of "super-fast" type. *b*, positive-working presensitized lithographic plate. *c*, Kalvar (vesicular) film.

at the image areas which would improve visibility under diffuse lighting conditions. Thus there are indications of wider applications in the future of a process which has the advantage of extreme simplicity in processing without external reagents; exposure and processing can take place in such close sequence as to be almost instantaneous.

When the bubble-image is projected, the image is black, i.e. a negative of the original. This may be desirable in its present major field of use: duplication of microfilm. But the projected image is seen as black because the bubbles are reflecting light away from its original beam path. Therefore if the image is viewed from the direction into which the light is reflected, it will appear white. By coating

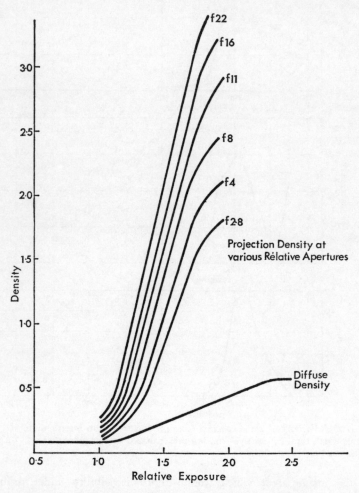

Dependance of vesicular image density on illumination conditions.

the vesicular layer on to a black support, the exposed parts will in normal viewing appear white, while the unexposed parts being transparent will appear black. Thus a positive print is possible from a negative-working system.

166

Diazo-metal systems

Another interesting group of negative-working systems based on diazo compounds involves the use of suitable metal salts in the sensitive layer. The action of light during exposure (see page 149) is to convert the diazonium compound to a hydroxy compound, which in turn can act as a reducing agent to convert a metal salt to metal (this type of reaction being the basic process of photographic development). The metal salt could be silver nitrate, in which case a silver image results. The N. V. Phillips Gloeilampen-fabrieken of Holland have used this type of system with mercury salts. The image of finely-divided mercury is then also subjected to the photographic process of *physical* development. This is by treatment in a solution which deposits further amounts of metal on to the initial mercury image. The principal attraction of this approach appears to be microfilming, on account of very high resolution obtained. An amplification factor (see page 179) of about 200 applies to this type of system (see also page 101).

Diazo compounds in photomechanical systems

Diazo-sensitizing was soon considered as a possible alternative to dichromates in the photomechanical applications involving hardening action on colloids. While it is true that colloid-tanning can be achieved by certain types of diazo compound, this approach was not particularly encouraging and gave way to systems related more specifically to diazo chemistry than to colloid properties.

One successful system utilizing colloid-tanning by diazo compounds is the Transfer Foil operating on the Van der Grinten reflex-printing principle. Transfer Foil carries on its lower surface a large number of indentations which are filled with a black composition and provide a screen with numerous clear openings. The whole of the lower surface is coated with a diazo-colloid layer. When unexposed to light, the colloid layer can become adhesive if moistened with water. After exposure, it loses this property. The screen and colloid layer are in contact with the original during exposure (see page 211). After exposure, the Transfer Foil is moistened, pressed on to a receiving surface, and the two promptly separated. The black composition transfers to the receiving surface where there has been no reflex exposure of the colloid layer. The receiving surface may be a white paper, transparent paper or film, or a lithographic surface. Thus single copies are obtainable, or a master for diazo duplication, or a lithographic plate for a large number of copies.

With diazo systems, complete independence is achieved of the dark reaction of dichromated colloids and of atmospheric humidity variations which cause difficulties such as in running albumen plates. Processes based on diazo compound behaviour made possible a break from the 'do-it-yourself' attitude in plate-making with the introduction of pre-sensitized plates. This is a contribution to the field of photoreproduction as important and far-reaching as was the invention of the photographic dry plate in substitution for the collodion wet-plate.

Although the principle of dye-coupling may be used to create the ink-accepting image of a lithographic surface, practical products have tended to rely on other properties which require more complex diazo structures than serve the purpose of producing paper prints. The term 'diazo-resin' has been coined to cover these structures, because they tend to resemble in molecular size and behaviour the substances used for polymerization into synthetic resins.

Formaldehyde can convert water-soluble gelatin to a water-insoluble complex; and form urea-formaldehyde resins, by linking together urea molecules. It will also combine with certain diazo compounds to give a complex structure: a sort of network of molecules of diazo compound joined together by bridges or links derived from the formaldehyde molecule. If this process is stopped before it has gone too far, the new compound retains water-solubility. Exposure of such a complex to actinic light converts its diazo groups to hydroxyl (phenolic) groups and this change presumably encourages immediate additional internal linkages to take place within the already large molecule. Possibly some of the molecules will also link together by a similar mechanism. Whatever the explanation, the net result is loss of water solubility and eventually an active 'dislike' for water. This condition is referred to as being hydrophobic or 'water rejecting'; the same condition may be described as *oleophilic* or 'oil accepting'. Thus a new approach to the preparation of photo-lithographic printing masters became possible. If of this type, the plate is essentially negative working. Processing can be very simple – in some cases merely washing away the unreacted diazo-formaldehyde compound with water.

Positive-working lithographic plates

A quite different group of diazo compounds which can be used for such purposes are derivatives of naphthalene rather than benzene, and have 'twinned' benzene rings in their structure:

Naphthalene
$C_{10}H_8$

This substance is familiar as the domestic 'moth ball', now to a large extent displaced by 'PDCB' (para-dichloro-benzene) for that purpose.

Simple compounds derived from naphthalene were used in early diazo copying papers relying on dye-coupling. The possibility of use in photomechanical applications only became apparent from the work of the German chemist, O. Süs. He determined the nature of the substances formed from these naphthalene-based diazo compounds under the action of actinic light. He found that the naphthalene 'double ring' was changed and a completely new type of compound formed:

COOH

$C_{10}H_8O_2$

As the new substance is an acid, it is soluble in water containing alkali, insoluble in water containing acid. Thus the exposed parts can be removed and the unexposed left as image. Moreover, by attaching large groups to the original naphthalene nucleus, the image acquires oleophilic properties. Such a substance, coated on to a lithographic surface, requires only the one processing stage of dissolving away the light-reacted parts to give a lithographic master. In this case, the plate produces a positive from a positive, whereas with dichromated systems, more complex 'reversal-processing' as with the deep-etch plate is necessary if working from positives. In this way the first truly positive-working lithographic system was obtained (1946). Similar, negative-working, plates followed the positive-working and are described later (page 174).

The simple system outlined functioned satisfactorily in very thin layers, but for good press performance (number of copies) the image has to be fairly thick to withstand the wear. Diazo compounds of a more distinctly resinous type are required, so that in practice the image is a diazo/resin complex, tough and oleophilic, and having the property of transformation to an alkali-soluble compound by exposure to actinic radiation.

Image reinforcement by lacquer and baking

In photolithography, it is a normal technique to strengthen an original image by a lacquering process. Resinous material in a suitable solvent is spread over the plate surface and attaches itself to the image proper. Thus more copies can be obtained by reinforcing the image once or more times during the run. This does of course mean stopping the press, cleaning off, lacquering and re-starting. With the diazo-based lithographic sensitizers it was found possible to incorporate the lacquer components in the original layer. The lacquer resins seem to associate themselves with the diazo-resin molecule, with considerable advantages in the processed plate and avoiding the lacquering stage in plate preparation.

A correctly-exposed positive diazo-sensitized plate is processed with one simple aqueous solution, the sole purpose of which is to dissolve away all parts of the layer which have undergone photochemical change. Processing must be complete and cannot be carried too far, whereas the lacquering operation requires a certain skill. Thus the modern diazo positive-working plate starts its life one stage further on than the earlier plates and gives a 'lacquered' image without a lacquering operation. Another advantage of this approach is that there are no residual solvents from a lacquer, which would soften the image if not removed by warming, etc.

These modern plates are usually regarded as providing some 30,000 to 50,000 copies under average handling. With good operation and a press in good mechanical order, 80,000 or more are frequently attained. Compared with a deep-etch plate performance of 100,000 to 250,000 the diazo product left something to be desired, though it is feasible to make say two presensitized plates in lieu of one deep-etch and show economies when producing 100,000 or more.

Recent improvements in the diazo positive plate have considerably lifted the anticipated length of run. By a baking process, a plate safe for 40,000 under prevailing conditions becomes satisfactory for 400,000. In this method, the fully-processed plate is baked for ten minutes in an oven at 240°C. The image undergoes a quite remarkable transformation, becoming unaffected by the chemical correctors which can normally be used, and impervious to solvent action. This makes the positive diazo-sensitized plate acceptable on Web-offset where heat-set inks are frequently encountered (for which reason negative-working photo-polymer types of plate are often associated with web-offset presses).

Web-coated anodized plates

The plates so far described use brush-grained aluminium as the base metal. The surface acquires by this process a fine grain which interferes less with definition in the image than the majority of ball-grained plates as hitherto known. The metal, in this 'scoured' and very clean condition, has a strong attraction for the diazo compound which therefore is tightly bonded without relying on a binder substance.

Similar sensitizers applied to electrolytically oxidized aluminium show even greater promise in terms of run length, without losing any of the basic attractions of the diazo sensitizer. This type of positive-working plate is also finding application in web-offset work.

Diazo presensitized aluminium plates were initially produced by web-coating, i.e. in a continuous strip of flexible material, going from the parent reel through the coating plant and then being re-reeled. Until recently, anodizing plant was not available capable of producing a continuous web of anodized aluminium; hence the adoption of brush-graining in the first place. Even now, continuous anodizing of a web is limited in web width. A second limitation in web-coating is the thickness of metal, which must of course negotiate various rollers in the coating-machine and cannot be turned through too acute an angle. Metal of 0.005in. to 0.012in. has commonly been used, and for the largest sizes 0.015in. This may be compared with the 0.025in. normally used by the printer, though by repeated re-use this comes down to about 0.012in. before it is discarded. The two requirements of an anodized surface and metal of gauge greater than 0.015in., effectively prevented further extension of web coating until very recently. The large-scale manufacture of presensitized metal in sheets was possible and plates of this type have become commercially successful.

Thus a comparatively short time – some 25 years – has seen the complete evolution of diazo-sensitized lithographic plates of highly sophisticated construction. These developments span the original laboratory work of Dr. Süs, its tentative application to the lithographic situation, significant improvements in conjunction with resin chemistry, extension of the image life by baking, and finally the conjunction with anodized metal; an achievement which has made possible the adoption of all-positive reproduction once any original negatives have been contacted to positives. This is of particular importance in connection with modern developments in photo-typesetting, in which an alphabet negative is used in a mechanical enlarger, to produce photo-set positive film in place of the traditional metal type in the 'galley'.

Bi-metal and tri-metal plates

Another extension of the basic chemistry involved is the sensitization of so-called 'bi-metal' or 'tri-metal' plates for lithography. Such plates are not likely to be encountered in reprography because they are required for runs of one million copies upwards. They rely for their printing image on two different metals, one of which is hydrophilic and the other oleophilic. Thus copper forms a printing image with good acceptance of greasy ink, whereas chromium or stainless steel does not.

A bi-metal plate might consist of stainless steel to which has been given a thin coating of copper by electroplating. It must be sensitized with a material capable of giving a suitable etching resist, because the copper is to be removed except where ink-acceptance is required. A negative-working sensitizer would require exposure under a negative in order to protect in etching the parts which are to print. A positive-working sensitizer could also function and allow exposure under a positive.

Other combinations of metals used in this way are chromium on copper and copper on aluminium.

In the bi-metal plate, one of the metals has to form the 'base' on to which the other is deposited electrolytically. In an effort to reduce costs, the tri-metal plate comprises a base of steel or zinc, on to which first one and then the other of the bi-metal combination are deposited. Thus zinc or steel may be electroplated first with copper and then with chromium.

As in each case copper is to form the printing image, and chromium, stainless steel or aluminium the non-printing areas, it is sometimes necessary to form a negative resist and at others, a positive resist. Also it may be necessary at times to work from positives and at others from negatives.

Etching resists from diazo compounds

Diazo compounds are capable of producing positive-working etch resists suitable in these two lithographic applications, though the requirements differ somewhat from the straight provision of the oleophilic litho image. There is the same dependence upon preferential re-solution of the exposed parts of the layer. The 'developing' solution which achieves this is still an aqueous solution without solvent, etc., hazards. A similar positive-working sensitizer has been applied to copper and zinc sheets for photoengraving. Diazo chemistry thus shows as much versatility as the dichromate/colloid family but with the various advantages which have been indicated.

Traditional photo-engraving for line blocks is a lengthy and costly business, more complicated than any of the lithographic plate-making processes. A new factor in block-making is the 'powderless-etch' process introduced in 1953 by the Dow Company of America (hence, Dowetch and 'one-stage' process, as will be seen from what follows). In powderless etching the photo-resist image is formed by the methods already described, but etching is carried out in a machine which throws a spray of etch liquid on to the plate while the plate itself is rotated. The etching bath contains, as well as the actual acid or salts for dissolving the unwanted metal, a 'protecting agent' which serves the same purpose as the fused resin powder of the dusting-up process: it prevents the etch bath from acting laterally and undercutting the surface parts protected by the resist. Rate of etching is under control and there are none of the delays of traditional photo-engraving. The net result is that the making of blocks for letterpress printing has been brought nearer to the preparation of lithographic plates in process and time costs. The success of diazo-based resist coatings has also given to the process engraver the advantages of pre-sensitized metal. Positive-working resists are attractive where screened positives are produced, or photo-typeset text positives, and allow time-saving multi-exposure techniques.

Thus apart from its original and continuing use for copying translucent drawings, typed masters, etc., the diazo processes have taken on an immediate significance in the changes now affecting the printing industry at large, principally by virtue of making truly positive-working systems available.

Positive-working etching resists have also extended into the fields of printed circuitry and photo-fabrication (chemical milling). These, again, initially grew up around negative-working systems of photo-polymer type. The techniques of preparing the design to be etched are well established and it cannot be expected that these will change rapidly. However, the much higher resolution possible with positive-working diazo-sensitized layers has attracted attention in certain areas where maximum fidelity is essential (for example, optical diffraction gratings). Photo-fabrication is more fully described in the section dealing with photo-polymer systems (see page 182).

Negative-working diazo presensitized plates

While the major contribution of diazo chemistry in photomechanical work is in positive-working systems, there remains a considerable field of interest in negative-working sensitizers which can be used for pre-sensitizing and avoid the disadvantages of dichromated colloids. In the simple situation described above to explain the function of the diazo resin in working positively, imaging relied upon exposure to convert the sensitizer into an acidic substance which could then be removed by an alkaline solution. However, if the sensitizer is constructed with sufficiently basic substituents the effect of actinic radiation is again to change the double benzene ring structure of the naphthalene nucleus; but this product can then form an 'inner salt' between its original basic parts and the newly-formed acidic part. After exposure an acidic solution is used to dissolve away unchanged sensitizer and to leave the new compound as a negative image of the original design used in exposure.

An alternative approach to a negative-working plate is with simpler, benzene-based diazo sensitizers containing somewhat reactive substituent groups. The effect of exposure is to split off nitrogen, leaving more highly reactive substances which polymerize, i.e. react together to form larger molecules. These are less soluble in organic solvents, dilute acid or dilute alkali and it is not difficult to find a developing solution which will leave the light-produced polymer as a negative image and expose the base which is hydrophilic.

Dry lithography

On page 83 the term 'dry offset' was used to describe the printing method whereby a raised printing surface was used in the situation most usually found with offset lithography; that is, inking of the raised surface, transfer of the ink to the blanket, followed by transfer from the blanket to the work surface. In this way the advantages of offset printing were obtained but without the problems of ink/water balance, or of image wear, which can arise in offset lithography. This method therefore offers a bridge between letterpress and planographic printing and prepared the way for printing presses which could operate from relief or flat printing surfaces, on the offset principle, with or without a dampening system according to need.

A further development of possibly far-reaching effect is the 'Driographic' process launched in the USA during 1970 by the 3M Company. In this, a truly lithographic printing surface is involved, but the plate is run dry. The plate does not require to be kept in a wet condition for it to differentiate between image and non-image areas. Thus it is run on conventional offset-litho presses but with the whole dampening system made inoperative. Plates of the type described reverse the original situation in that the surface is oleophilic, and it is the non-printing areas which have to be created. The method is for most purposes therefore better adapted to photo-imaging than to hand-drawing.

The principle of Driography is to have a plate surface of cleaned and slightly oxidized aluminium, but not prepared as by anodizing which creates a permanently hydrophilic surface. In this plain condition, the metal is oleophilic and readily accepts printing ink. The whole of the plate is covered with a sensitized layer of a silicone gum or silicone rubber, this layer also containing a diazo compound as sensitizer. The effect of exposure is to make the silicone layer removable by a solvent mixture, whereas the non-exposed parts remain intact on the plate. After this processing, a water-wash, and drying, the plate is ready for the press. The silicone rubber is formulated to give a surface which even in the dry state, is ink-repellant. It would appear also that the construction of the layer(s) on the plate must give the correct degree of adhesion of the silicone material to the metal, as an aid to successful processing. As described, the Driographic plate is negative-working.

While it may be possible to use photosensitive materials other than diazo compounds to achieve an imaging method for this purpose, it seems that the Driographic plate as marketed does rely on a diazo substance – further evidence of the flexibility and adaptability of diazo chemistry in the photomechanical context. The plates are of course offered presensitized. Initially the main use of Driographic plates has been in the printing of business forms in relatively narrow widths, but plates up to 60in. wide are expected, in gauges ranging

175

from 0.005in. to 0.012in. Since there is no ink/water balance to be established on the press, printing commences from the second or third copy. It has been necessary to formulate inks for the purpose, but this is hardly surprising in view of the complete departure from the normal lithographic situation. There are problems still to be overcome; one is created by the very advantage of running dry, for this encourages paper dust and fibre to accumulate on the blanket, with the possibility of eventual damage to the plate by scratching. Whatever the remaining problems, Driography stands as a most interesting development capable of considerable impact on the printing and reprographic scene.

Early history of photo-reproduction by diazo compounds

1858 – discovery of diazo compounds by Peter Griess.
1864 – first use of dye-coupling reaction by Griess to produce an azo dyestuff.
1881 – light-sensitivity of diazo compounds recognized by Berthelot and Vieille.
1884 – first attempt at a photo-process by Dr West.
1889 – second photo-process by A. Feer. (Both West and Feer produced negative-working systems.)
1890 – principle of positive-working established by Green, Cross and Bevan.
1895 – negative-working process by Dr M. Andressen.
1899 – negative process by M. Schoën, rather similar to that of Dr West and of Andressen.
1901 – process of O. Ruff and V. Stein.
1920 – positive process of Kögel, diazo compound and coupling component both present in layer (2-component system using diazo anhydrides and ammonia processing).
1925 – stabilized diazonium salts used in Kögel-type process by Kalle.
1927 – basic structure of modern diazo compounds disclosed by Van der Grinten.
1927 – one-component system and moist processing by Van der Grinten.

CHAPTER 11

PHOTOPOLYMER SYSTEMS

Photopolymers are the most recent addition to the main processes available for photoreproduction. They date from 1945 (W. E. F. Gates' British patent) although the requisite chemical knowledge accumulated over the previous 20–30 years.

Principles of polymerization

Polymers are complex organic substances involving long chains or rings of carbon atoms, with or without 'cross-linkages'. In structure they have some resemblance to the gelatin molecule already described. The basic unit of a polymer is a 'monomer'. Monomers are quite simple compounds which may be gaseous, liquid or solid under ordinary conditions. Under an appropriate stimulus, monomer molecules join together and produce progressively larger molecules with corresponding changes in properties: the larger the molecule for instance the lower its solubility in whatever solvents dissolve the monomer. A liquid monomer might be transformed into a solid substance of low melting-point and fairly high solubility, and this in turn become a solid of higher melting-point and complete insolubility in one solvent. The process of polymerization is normally induced by a chemical catalyst and the degree of polymerization is controlled by time, temperature and concentration of catalyst. The main industrial use of polymerization chemistry is in the plastics or synthetic resin industry.

A certain number of monomeric substances can also be induced to polymerize under the excitation brought about by absorbed radiation. The mechanism is that when one molecule of monomer absorbs its quantum of radiation energy, it achieves a higher energy state in which it can combine with a second molecule of the monomer. In doing so, the energy which activated the first molecule is passed on to the second which can then unite or combine with a third, and so on. In the simplest case, a long chain or linear polymer is eventually formed. By having a mixture of two types of monomeric substance, more complex structures are finally achieved.

Image formation

In undergoing these changes, there is no image formation in the sense of a visible colour change. The 'image' is a pattern in the monomer layer, having different physical properties from the monomer. Of these the most important is a change in solubility; by proper selection of the solvent used, it is possible to differentiate between the parts which were exposed to the radiation and those which were not. Thus, the majority of photopolymer systems involve a liquid processing stage after exposure, and are inherently negative-working.

If the monomer layer were additionally pigmented, then on dissolving away the unexposed parts a visible negative image would be obtained, just as in the case of pigmented dichromate/gelatin. However, such an application would not have much to recommend it over the well-established reproduction methods for single copies.

The main application of photopolymers has been in photomechanical situations. They are encountered in a variety of presensitized negative-working lithographic plates, as sensitizers for printed circuitry or metal engraving, and as relief printing plates. Very recently an ingenious use of photopolymers has been launched in America which avoids liquid processing, utilizing a change in adhesive properties in a peel-apart sandwich of two films.

Before describing these in greater detail, it may be pointed out that a successful monomer-polymer system for photoreproduction is somewhat limited in the substances available as monomers. For instance, a liquid monomer could be spread on, say, a glass plate and exposed in this condition; the effect of exposure is to convert the liquid to solid. Obviously such a system would be very limited in application and would involve a number of practical difficulties.

To be of general value, the monomeric substance should itself be solid at room temperature, so that it can be coated and dried on the support and then be handled without risk of damage. As the sensitive layer has this initial requirement, the changes which can take place on exposure to suitable rediation will not be so extreme as in the transition from a liquid to a solid condition. For this reason practical photopolymer systems tend to require choice of a suitable solvent and careful manipulation in processing after exposure.

178

Quantum efficiency and amplification

In considering light as radiant energy and the basic mechanism of photoreproduction, we have encountered the quantum as the energy unit which can, at most, affect one molecule of the sensitive substance absorbing the radiation. We have further dealt with the idea of 'process efficiency' as a combination of quantum efficiency and amplification of the primary photo change by subsequent chemical, etc., methods (notably, in the development of silver halide materials). The possibilities of high quantum efficiency in photopolymerization have attracted much attention in recent years; but some caution is necessary before deciding that photopolymerization is on that account a 'better' or more efficient method than another.

In the case of a photopolymeric system, the effect of absorption of one quantum of energy can only be to bring about a change or 'excitation' of one molecule. However, the process whereby one excited monomer molecule can link to a second, the second then acquiring the excitation energy and linking to a third, the third to a fourth and so on, means that photopolymer systems have a kind of built-in amplification property. While this is true in terms of the total number of monomer molecules involved in the change initiated by absorption of one quantum, it does not put the photopolymer systems on a similar footing to silver processes because there is no visible colour change as a result of the polymerization process. Hence it cannot be assumed that photopolymer systems are necessarily 'more efficient' merely on the grounds of a higher quantum efficiency. In the applications mentioned – photolithography, metal-engraving, relief printing-plates – the end-product of the photoprocess is the means whereby the ultimate reproduction in pigmented ink is made. The same applies to the peel-apart product where the photopolymerizable layer is pigmented. The 'amplification' achieved in this way exceeds that normally expected of a photoreproduction material.

In *Photographic Systems for Engineers* (SPSE Washington, DC) the editors have been careful to warn against preoccupation with quantum efficiency unless there is also sufficient regard for the amplification factor in the process. In this publication a useful comparison between recording processes is given (see table overleaf).

As the factor for the diazo process is taken as one, this must refer to the fact that one diazo molecule at most can be affected per absorbed quantum; whereas with silver, millions of molecules of silver

Amplification factors for recording processes.

Process	Amplification factor
Diazo	1
Photopolymerization	100,000
Electrophotography	1 to 10 million
Silver halide with photographic development	100 million

halide are changed to metallic silver for each one affected by the absorption of a quantum of energy. At the same time, there is a considerable increase in the visible effect when each residual diazo molecule is coupled to form dyestuff. Thus there may well be a greater visible result from one molecule of diazo compound when dye-coupled, than from thousands of monomer molecules which join together in a photopolymerization material.

Bearing in mind also that all the four processes listed as well as dichromate systems, give photomechanical systems in which the final result is pigment deposited on to a paper or other surface, they might all then be said to achieve the identical amount of amplification in the overall process. For this reason the choice of a photoreproduction process for a given purpose must take into account many more factors than the apparent energy efficiency of the primary step in the sequence.

Photopolymers in practice

Photoreproduction systems relying on photopolymer types of reaction are not as simple as suggested above, though they function on the principle described: a joining-together of two or more molecules under the excitation of radiant energy to produce more complex molecules of low solubility. In the Gates' patent, a liquid monomer or mixture of monomers is contained in a glass cell made from two sheets of glass separated by a gasket. After exposure – for which four hours is suggested – the cell can be taken apart, the polymerized sheet removed and 'developed' with suitable solvent. This may take from half to several hours. The developed relief image is finally dried and can then be mounted as a printing 'block', or attached to a printing cylinder.

To make a more practical system, the unexposed layer must usually consist of more than simple monomeric substances. Imagine that instead of single beads spread over a tray, the beads were strung into short 'chains' of a dozen or so per chain; and that some of the chains had side-branches of several beads. Then it would only require a few inter-connections between the chains, either in line or at the branch-ends, to produce a big network of beads which would be 'all of a piece'. This would have been achieved much more easily than if the same network had to be built entirely from single beads.

Resists based on polyvinyl cinnamates

One of the widely used systems of this type has been extensively developed by the Eastman-Kodak company and relies on polyvinyl cinnamates. The basic compound in unexposed condition may be thought of as a long chain (the polyvinyl part) with side-branches from cinnamic acid. It is possible for two adjacent cinnamic acid molecules to join together under light action and to form a new four-membered ring (two from each cinnamic acid molecule). This action between the cinnamic acid branches on the vinyl chain rapidly achieves a large net-like structure of much reduced solubility. The system was originally envisaged for etching resists (the original KPR = Kodak Photo Resist) but is also applicable to photolithographic plates as the exposed parts have good ink acceptance. Layers of this type produce a tough film on exposure capable of withstanding fairly heavy wear and tear on the lithographic press; also because of their resistance to solvent action, they are satisfactory for use with 'heat-set' inks (which rely for drying on the evaporation of a solvent).

A simple system relying on polyvinyl cinnamate alone is not very rapid in undergoing the photochemical change described. This is due to the fact that it absorbs, and therefore is energized by, a narrow wavelength range around a peak of 300nm. The effective speed of the system can be very much increased by a variety of other organic substances added as 'sensitizers'. Some of these extend the wavelengths absorbed well into the visible region, and allow a sensitivity to be reached appreciably greater than that of typical dichromated colloids.

Such layers not only show good speed, but also possess great stability in the unexposed condition, considerable tolerance in the solvent processing stage and good resistance to acidic etching solutions as used in printing circuitry, etc.

From the purely chemical point of view, the behaviour of the poly-vinyl cinnamate type of substance on absorption of suitable radiation is described as 'cross-linking' rather than as photopolymerization in the general meaning of that term. The cross-linking reaction involves what are called 'double bonds' in the cinnamate structure and true polymerization is similar in this respect. Both types of system are quite different from the classical systems involving silver, iron, and chromium compounds and to a less extent from some of the diazo situations; for in all these previous cases, the continuing action of light on the primary substance depends upon the presence of another compound to 'take up'

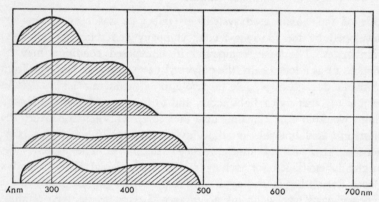

Effect of additions on spectral sensitivity of polyvinyl cinnamate.
Top to bottom: unsensitized; p-nitroaniline; 2,4-dinitroaniline; picramide; 3-methyl-1,3-diaza-1,9-benzanthrone. Robertson, van Densen & Minsk.
Journal of Applied Polymer Science **2**, 308 (1959).

the products formed and so keep things going. Systems as now described relying on 'double bonds', though energized by sensitizing additions, are more truly self-sufficient in that two molecules each carrying double bonds are able to combine together to form one new more complex molecule. In this way, the entire results of radiant energy absorption are built into the resultant image.

Photofabrication processes

The availability of the Kodak range of photo-resist preparations has extended the scope of photoreproduction to the relatively new concept of photofabrication. By using the techniques of drawing on a large scale, camera reduction to a negative of the final size required, and photosensitizing the material in question, it becomes possible to fabricate by chemical engraving, metal components which would be impossible by any other method. Intricate components can be made in thin-gauge sheet-metal with great accuracy and without the stresses or burrs which mechanical punching would produce.

Photofabrication methods are rapidly extending into many manufacturing fields with particular emphasis on electronics and miniaturized equipment. They are equally applicable to 'coarser' requirements such as name-plates and decorated metal articles. Where the purpose is the removal of relatively large amounts of unwanted metal, the terms 'photo-milling' or 'chemical milling' are used. This might, for example, be to reduce weight as in aircraft components. Where the pattern of the design is finer or the amount of removable material much less, the tendency is to refer to 'photo-etching'; printed circuitry and nameplates are typical. A newcomer to this field is 'photoelectroforming', where instead of using the resist as a protection in an etching process, the photo-image is formed on a conducting surface and becomes a stencil for the electro-deposition of metal. The deposited metal is subsequently stripped off and forms the required part: examples are fine screens and graticules for optical equipment.

Comparison of diazo and polymer resists

In the section on diazo compounds, reference was made to etching resists for photofabrication and a comparison with the photopolymer type is of interest. Both types share certain advantages over dichromated colloids (for instance, no dark reaction and therefore the coated material is storable before use). Photopolymers are negative-working and show good resistance to acids, alkalis and plating solutions. The diazo-based systems are positive- and negative-working and have a higher resolution than photopolymer types, but are limited in the etch and plating baths which they withstand. In processing, the photopolymer requires use of a solvent whereas the diazo layer uses an aqueous solution. To remove the photopolymer resist after use involves vigorous attack with a 'stripper' while the diazo image can be removed by re-exposure to light and applying the aqueous developer.

Relief printing plates from photopolymers

There are two types of photopolymer relief plate now available: the Dycril series by the Du Pont Company; and the nylon plate by Time, Inc. or Nyloprint by BASF. The chemistry of these products would not be suitable for discussion here. Two points to note, however, are

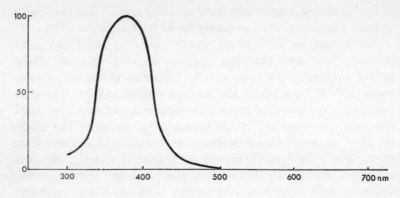

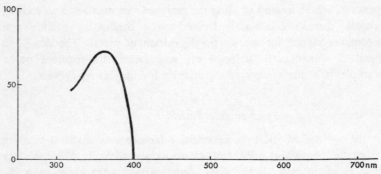

Top: photopolymerisation system: 9-vinyl carbazole and carbon tetrabromide. This system has a measured quantum efficiency of 7, but its speed is described as only "somewhat higher than commercial diazo films". *Photographic Science and Engineering* **14**, 97–100 (1970). *Bottom*: photosensitive nylon. The curve has been corrected to show the response to an equal energy distribution. Leekley, Sorensen, Byers *et al.*, *Relief printing plates from photosensitive nylon*, Technical Association of the Graphic Arts 9th Proceedings, 1957.

that the photopolymerization effect can be enhanced by incorporating an initiator or can be prevented by an inhibitor.

An initiator can be thought of as an energy carrier. While it is true that certain monomers will become excited by absorption of radiant energy and commence the process of polymerization in this condition, in practice there are not many compounds which show this ability. A suitable initiator absorbs the energy – probably over a wider wavelength range – and reaches an excited condition. The initiator has to be capable in its excited condition of holding this energy for a sufficient time to trigger off the required polymerization reaction proper.

Once started, it might be thought that polymerization would continue until all molecules present had been caught up and made part of one immense molecular network. However, there are various ways in which the process can be terminated, and one of the most decisive is the presence of oxygen. The system used for the Dycril plate is of a type in which photopolymerization does not take place in the presence of oxygen; the unexposed Dycril plate is therefore 'conditioned' in an atmosphere of carbon dioxide for 24 hours before use. Oxygen in the coated layer is thus displaced, that is, the inhibitor is temporarily removed from the sphere of operations.

The chemical system of the nylon plate is distinct from that of the Dycril, being based on polyamide, also using an initiator addition. Both types have similarities with respect to construction: a base sheet carries a bonding layer and on to this is coated the photopolymer system. The base sheet may be flexible: plastic, steel or aluminium; or rigid: aluminium. Overall thickness ranges between 0.4 and 1.5 mm. in flexible forms, 1.75 and 6.12 mm. in rigid. The relief image varies in depth from 0.2 to 1.0 mm.; thus the system produces an appreciable thickness of photopolymeric substance. The only processing after exposure is to dissolve away any unpolymerized (unexposed) parts of the layer. With Dycril this is done with a dilute alkali, with the nylon type alcohol or alcohol/water solutions are used. The Time, Inc. plate can be additionally hardened by heating. Preparation of these plates requires 20–30 minutes overall, using special processing equipment. Here again therefore, modern processes have given to letterpress printing much of the time-convenience of lithographic printing.

Du Pont sandwich-construction materials

Photopolymerization processes are essentially negative-working and normally require a wet processing. A most ingenious product has recently been introduced by the Du Pont Company under the name Crolux. This is a sandwich of two polyester films with the polymerizable layer in between. This layer is pigmented, normally black and of high actinic density. In the unexposed condition the middle layer has better adhesion to the bottom film than to the top one; i.e. if the top film is peeled off the layer would all adhere to the bottom film. The effect of exposure is to change the adhesive character; those parts of the layer which undergo polymerization also acquire a strong bond to the top sheet, so that when the two sheets are separated after exposure the top sheet carries away a negative image while the bottom sheet retains the unchanged material as a positive image. The sandwich construction prevents the inhibiting effect of oxygen on polymerization.

As described, the bottom sheet is 0.004in. thick and the top sheet 0.001in. Note that the layer has to receive the actinic radiation *through* the top sheet, and that this imposes a separation of at least 0.001in. between the original and the photo-sensitive layer. For best results it is therefore necessary to use 'critical' conditions in exposing (see page 329). Note that the positive image consists of the original layer material, which has not undergone polymerization. After separation of the two sheets, the positive image can be given a further exposure to 'cure' it, when it becomes less sensitive to accidental damage by scratching or solvent spillage. One user has said of Crolux that the new film did not fall below the quality of conventional diazo products in any of the tests carried out. This is a tribute to the diazo products as one of the attractions of the Crolux system is that by the pigmentation of the layer, a considerable amplification factor exists.

Photopolymeric printing inks

Somewhat akin to the type of pigmented, polymerizable layer of Crolux, though not used in a photo-imaging situation, is a new type of printing ink which appeared in the USA during 1970. This is 'Suncure', of the Sun Chemical Company. It is made without the usual solvents, and relies on photopolymerization to solidify the 'wet' printing ink on the printed article. This is brought about by exposure to the radiation of UV lamps, presumably similar to the tubular mercury-vapour discharge lamps already described. The cost of these inks is 25–50 per cent higher than conventional inks. Advantages are for example the avoidance of solvent vapour in the press-room and in the discharge to outside atmosphere; completely 'dry' printed sheets coming from the press; and avoidance of anti-offset powders with their complications. Without exposure to UV radiation, these inks remain soft indefinitely, which simplifies cleaning-up operations on the press and elsewhere.

A further development of the use of photopolymers was launched by Du Pont in 1968. This is in connection with printed circuitry and offers much of the attraction of a presensitized product without in fact being one! The photopolymeric layer intended to create the etching resist of the printed circuit is supplied as a sandwich between two sheets of polyester film. The sandwich is really a device making it possible to market the photopolymer, preformed as a layer, for transfer to the metal-clad board of the circuit. This is the technique of the gravure tissue, without the necessity of treating the tissue in a dichromate bath before use.

The Du Pont system (Riston) requires a laminator which strips one film off the sandwich and applies the remainder to the circuit board under controlled heat and pressure. The top, cover, sheet remains in place during exposure and is removed for development. Thus the photopolymer itself is fully protected from handling damage, or dirt, before and during use. Exposure of the layer has to be made through the cover sheet, which fact must impose some limitation on the acceptable exposure conditions (see page 198). After exposure of the layer to the design negative, the cover sheet is in turn peeled off and unpolymerized parts of the layer dissolved away with solvent (chlorinated hydrocarbon type, non-flam and low toxicity) at room temperature.

The laminator operates at 10 feet a minute and accepts boards up to one quarter inch thickness. One or both sides of the board can be laminated at the same time. At present, working width is 18in. but there is no inherent width restriction. This allows the system to be used in a flow-line keeping pace with other operations involved. The resist can be used with all normal etching and plating solutions, and after etching is removed with organic solvent. No spectral sensitivity curve has yet been published but high-pressure mercury vapour lamps are recommended for shortest times. Sensitivity would therefore appear to be mainly in the 365nm area.

CHAPTER 12

PHOTOCHROMIC SYSTEMS

The so-called photochromic materials are substances which undergo spontaneous changes in colour when suitable radiant energy is absorbed. The colour change is immediate and requires no further processing to be visible. At first sight this is an attractive situation; but the practical applications are somewhat limited, because the changes involved are reversible. Thus a colour-change might be brought about by radiation of wavelength 400nm and below – that is, beyond the visible blue region. The substance undergoes a structural change in order to produce the colour change. But when the new coloured substance absorbs radiation of another wavelength, perhaps in the visible region, the structure reverts to the original uncoloured form. This type of change presumably occurs in the process of vision in the retina of the eye; the photochromic systems appear to have their place in a similar context – to provide an immediate result for a given purpose, then revert to the original condition and a state of readiness for further action.

This in fact is the case in the one practical application of a photochromic material so far known, the PCMI technique of the National Cash Register Company. This is a miniaturization system in which a photochromic material is used to 'observe' each micro-image frame as it is formed. Should the image be faulty in any respect, it can be erased and re-recorded when the fault has been remedied. In this way several thousand separate images are built up on a single piece of photochromic material measuring some 4×6in. This completed matrix is then used to prepare permanent same-size copies on a high-resolution photographic film. Once these copies have been made, the matrix is completely erased for re-use. Thus the reversible change which is characteristic of a photochromic system is used to advantage in a situation with a specific requirement.

As might be expected, the substances which undergo this type of change are fairly complex; the changes themselves are mainly of interest to the chemist, especially in considering possible ways of fixing the image. Such a method would allow photochromics to enter the field of general copying methods. In recent years fixation processes have been described and various novel ways of using such techniques mentioned.

Fixation by 'dry' (gas) and 'wet' (solution) processing is possible, but it would seem that these methods of producing a permanent image remove the chief attraction of such systems, at least so far as conventional copying is concerned. No doubt as time goes by other situations will arise in which the properties of photochromics make them especially attractive.

Photochromism and colour vision

Although the photochromic substances and their various possibilities are of great technical interest, they do not as yet come into the general reprographic sphere as operating processes for photoreproduction. As, however, the visual process itself depends upon this sort of reversible photochemical change, the behaviour of this type of substance is of more than passing interest. In the section dealing with vision (pages 13–19) there is reference to the 'visual purple' pigment, or rhodopsin, of the retinal rods which are responsible for the ability to see at low illumination levels – the 'dark-adapted' condition. The spectral sensitivity of visual purple, measured by actual extraction of the pigment, is shown on page 190.

In the earlier section (page 17) there is also brief reference to the three colour stimuli to which the cones respond as the mechanism of normal colour vision. This explanation of colour vision, until very recently, amounted to a *theory* supported by the experimental proof that colour sensations of almost every kind could be reproduced (synthesized) from correctly proportioned mixtures of three basic radiations or wavebands. As yet there has been no extraction of the three substances from a retina, which could be shown to have the spectral response required by colour-vision theory. But it has been possible for measurements to be made on single cones of a human retina, using a microspectrophotometer. The curves c, d and e page 190 have been obtained in this way, and for the first time give direct evidence of the nature of colour vision, confirming the long-held theory first formulated by Young (1773–1829). It is still necessary to view these results with some caution, but it can now be accepted that three separate colour 'sensing' substances do exist in the cones of the retina.

As with rhodopsin, these three substances are photochromic, in the

189

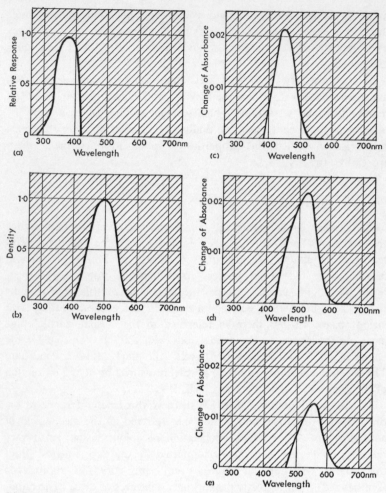

(a) Spectral response of a photochromic material, from "Photographic Systems for Engineers", p. 88, ed. by Brown, Hall & Kosar. Society of Photographic Scientists and Engineers. (b) absorption curve for pure visual purple, R. J. Lythgoe, *J. Physiol.* **89**, 1937. (c) to (e) three types of colour receptor in human cones – blue, green and red respectively. From Brown and Wald, *Science* **144**, 45–52, No. 3, 614, 1964.

sense that they undergo reversible change, thus making the photo-sensitive system self-regenerating. Photochromics used in reprography do show fatigue after a time but the eye, as a living organism, can also carry out running repairs so that a person in reasonably good health can overcome temporary fatigue and restore full operational conditions. Now that the immense subtlety of the visual processes is becoming clearer, we should be careful never to be too emphatic in our personal judgment whenever colour is involved, unless we are in good health and able to make the judgment under suitable viewing conditions which minimise fatigue; or wait until the normal situation is restored after being fatigued.

The three curves are referred to as *difference* spectra, because the measurement method records at each wavelength the difference betwen absorption in the dark (rested) condition and that after exposure to a flash of yellow light. The three curves have maximum absorptions at about 450 mm (blue receptor), 525 mm (green receptor), 555 mm (red receptor). The measurements were made with the light passing through the cone axially, i.e. in the direction which is normal in the living eye.

CHAPTER 13

PHOTO-PRINTING METHODS

We have seen that the excitation of atoms or molecules by heat or electricity can give rise to radiant energy; and that suitable radiation can be absorbed by sensitive substances, the absorbed energy bringing about various types of molecular change. It remains to consider how these two facts are brought together in the practical situation of producing a photoreproduction.

The term radiant energy was used to emphasize one of the distinguishing features of electromagnetic energy – the ability to act at a distance. It also can traverse a vacuum, unimpeded. But when passing through gases, certain wavelengths of the radiation may be absorbed by the substances present in the gas – as with the Fraunhofer lines of the sun's radiation (page 49). Most reprographic operations are carried out in the ordinary atmosphere and scattering of the radiation can occur through dust or water droplets forming mist or fog (page 50). These conditions cause problems in photography especially over large distances, but not in reprography.

The radiation pattern

Assuming then that conditions are suitable, photoreproduction functions by exposing the sensitive material to the radiation in a controlled fashion so that the required image can be recorded. This requires that the radiation reaches the receiving surface in a *pattern* related to the object being copied. The creation of this pattern requires the use of materials which may themselves bring new problems through selective absorption of certain parts of the radiation, or by affecting some wavelengths differently from others.

There are two ways of forming the pattern of radiant energy which is to be recorded. In the *optical* method, a lens system is used to focus light from the original object on to the sensitive material. In *contact* work, a shadow of the original must be cast on to the sensitive surface. Contact work may be by either direct transmission, or by

192

reflection from the surface of the object. Reflection or reflex printing is a special situation limited in the main to certain silver-sensitized materials and applicable to diazo products under very limited conditions. It remains a contact method, with the same basic considerations as transmission printing.

Function of optical systems

Apart from the camera, optical systems are used in the enlarger (projection printing), the telescope and microscope, and the episcope (often used in electrophotography). Clearly one of the limitations imposed by any lens is that its size restricts the amount of light it can receive and transmit to the sensitive material. The importance of aperture in the lens is relative to the focal length of the lens, hence the practice of expressing aperture as an f-number which is aperture diameter in relation to the focal length. When a lens is used in some types of projection printer, it becomes possible to build an elaborate supplementary system of mirrors and lenses so as to pack more light through the system and thus reduce the amount of time required to produce a given photo-effect. This is why although in the ordinary way camera photography has to rely on silver-sensitized materials (making use of the amplification effect of development) it is possible to make a projection printer to work with diazo-sensitized products. By and large, however, it is silver processes which are naturally geared to optical methods, at least in camera work; though electrophotography with its amplification by toner is also 'camera-speed', even if normally limited to print-making rather than camera use.

We spoke above of the lens focusing light on to the sensitive material. The energy pattern created by the lens action is referred to as an *image*, but this name is also given to the final visible effect within the sensitive layer. For the moment, think of the image solely as the pattern of light or other radiant energy which is to be 'captured' by the reproduction process. For example, a lens as used in a telescope, microscope or camera forms an image by bringing together in one plane the energy which the lens has received from the object and has transmitted. The image is said to be *in* the focal plane of the lens and can be seen as it were in mid-air by focusing the eye on this focal plane. (Technically, it is only 'real' images which can be seen, but it is this kind of image with which we are concerned.) To make it easier to see the image, it is normal to place a sheet of ground glass in the focal plane of the camera lens, or an eyepiece in the telescope or microscope, but the image is there and visible without these devices.

Exposing the sensitized surface

The reproduction or copy of the original is brought about by allowing the image to fall on the sensitized surface for a sufficient time to bring about the necessary chemical changes. This is the actual exposure operation. Exposure must continue as long as necessary to achieve the chemical changes, hence we must recognize the importance of the *intensity* of the radiation reaching the sensitive material, and the *time* for which exposure must be continued. *Roughly speaking*, halving the intensity would require double the time. Normally, it is necessary during the exposure operation to protect the sensitive material from all radiation other than that of the image being reproduced. Exposure must be thought of, not only as allowing radiation to reach the material in some places but also preventing it from doing so in others. The avoidance of 'stray' radiation is an important part of the photo-reproduction process.

Contact printing by transmission is the method by which most plan and document copying is carried out. When an object is brightly lit, a corresponding shadow is formed which reproduces the shape of the object. To retain the most faithful shadow-shape, it must be cast on a surface which is parallel to the object itself; otherwise some degree of distortion occurs. This shadow—or rather, the radiation whose absence creates the shadow – is an energy pattern within the context of photoreproduction operations. Usually it is necessary for the original and the sensitive material to be in close physical contact to achieve the condition that radiation reaches some areas of the sensitive material but not others. However, the term 'contact' is a relative one and we shall see that there are degrees of contact which in conjunction with various types of radiation form the total environment of the exposure operation. In distinction from contact printing, the optical methods operate by projecting an image from a distance at the expense of the low light-transmission of a lens.

In optical printing, there may be only a lens between radiation source and sensitive material; in contact printing there may have to be a sheet supporting the original and a sheet of glass, etc., to sustain the contact condition. Therefore we must consider what effect there may be on the radiation between its leaving the source and reaching the sensitive layer. These effects can be anticipated from a knowledge of the spectral absorptions of glass, plastic sheets, etc., but it is more meaningful to deal with what is *transmitted* rather than what is *absorbed*, because it is this which will carry into effect the photo-process.

Factors arising from light transmission

When first considering the properties of light as a form of energy, differences in the behaviour of materials in transmitting energy were illustrated. We should now attempt a more accurate picture of the situation.

As with the energy conversion processes of the various light sources, no energy is lost or disappears; all can be accounted for. In the same way the total radiant energy falling on lens or original can be accounted for. In the case of a completely black material no energy is transmitted, and there is 100 per cent absorption, in the visible regions. The absorbed energy will for the most part re-appear as heat – the temperature of the material will rise.

Total or 100 per cent absorption indicates complete opacity. The reverse of opacity is to be transparent, which implies nil absorption and 100 per cent transmission. Complete transparency is not met with in practice and the word is used commonly not in its precise sense, but to indicate clarity of vision and absence of diffusion. Those materials which fall between opacity and transparency are translucent and it is the translucency of materials which is significant in reprography. A translucent material is said to transmit light with diffusion.

One normally speaks of polished glass, or of a clear plastic film with polished surfaces, as being transparent; but if a piece of clear plastic film is laid on, say, diazo-type paper and exposed to actinic radiation, it will be found that when exposure has just destroyed all diazo compound in the margins unprotected by the film, a sufficient amount remains beneath the film to give a slight background image. It might require a 10 per cent longer exposure before the diazo compound lying under the film is also destroyed, because the film is absorbing – failing to transmit – some 10 per cent of the incident radiation energy which is actinic to the diazo compound.

Part of the lost 10 per cent will have been reflected back from the front surface of the film, part absorbed by the molecules comprising the plastic sheet to re-appear as heat, and part will have been scattered on reaching the back surface. If instead of clear film a sheet with one matted side is used, some 25 per cent longer exposure might be necessary, in which case the lower transmission is attributable to the increased scatter which takes place at the matt surface.

The percentage transmission of papers is appreciably lower than that of plastic sheets, on account of their fibrous structure and the presence of sizing to bind the fibres together. A detail paper might transmit 40–50 per cent of the actinic radiation, other papers progressively less. The transmission of very white papers may have been lowered by the inclusion of the white pigment titanium oxide which has a high absorption of ultra-violet light, or by the use of so-called optical bleaches or brighteners. The function of an optical bleach is to fluoresce, i.e. to absorb radiation to which the eye is insensitive and to re-emit this energy in the visible region (hence the ultimate accuracy of the phrase 'whiter than white').

Choosing the working conditions

The approximate figures mentioned are easily observable in terms of the exposure time increase to give total bleach-out. To employ photo-reproduction processes to the best advantage it is necessary to understand not only the total quantitative effect of the translucent materials, but also the changes in quality which may take place as between the incident radiation and the transmitted radiation. This is seen as change in the spectral character of the two.

With optical methods the principal source of such effects is obviously the lens or other optical components. By and large the transmission of a lens system is not high in the ultra-violet region. Enlargers offer more scope than do cameras, because there is easier control of the radiation source and a high proportion of the total radiation can be 'gathered up' in the illumination system. Nevertheless, optical systems are also prone to losses and defects which arise from surface reflections and scatter. Modern lenses are 'bloomed' to reduce these effects.

Apart from choice of sensitive material and of radiation 'source', each method requires care to ensure best results – using the word 'best' to mean the closest approximation in the copy to the original. The lens requires that the image produced at the photosensitive surface be in focus; contact printing at its best requires that the original and the sensitive material are literally in contact. In practice, neither condition is completely realized but it is not difficult to obtain copies which are satisfactory *for the intended purpose*. The quality of a reproduction must be assessed in terms of its suitability for the end-purpose. There is no point in applying the criteria of excellence if the copy will be thrown away as soon as it has been read; but it is necessary to strive for perfection where the end-result must serve some critical purpose.

A first requirement when using either optical or contact methods is that the radiation should be uniformly distributed over the area of the work. The camera copy-board must be lighted as evenly as possible. Completely uniform lighting is not achieved but there is sufficient working latitude to minimize the effect of uneven lighting to the point of insignificance; in the case of silver-sensitized materials the factor of chemical development of a latent image is helpful in this respect. In contact printing the requirement for uniform radiation conditions is probably greater, because most photoreproduction layers rely solely on the absorbed radiation to bring about the total photo-effect. There is more latitude when exposing negative-working materials than positive, or at least a greater tolerance of over-exposure.

Depth of focus and depth of field

After uniformity in lighting arrangements, accuracy of focus and intimacy of contact are the next factors to take into account. No lens is perfect in all its aspects; some aberrations can be corrected only at the expense of introducing others. Corrections are, for instance, for flatness of field or for colour (because each visible colour and the invisible radiation which the lens transmits, all come to a focus at their own, slightly different positions).

Lens design ensures that when the lens is in the position of best focus, the image produced in the plane where the sensitive material is located (the focal plane) is as free from defects as possible; generally speaking the more complex the lens the better its performance and the higher its price. But there is a working distance either side of the focal plane in which the lens will still produce an image of a satisfactory quality, i.e. equally sharp for practical purposes. This working distance for acceptable results is referred to as depth of *focus*, and there is a similar depth of *field* in relation to the position of the original or object to be photographed. Depth of focus is only of interest where focusing can be effected by altering the position of the sensitive material. The usual method, however, is to move the lens or copy so as to increase or decrease the lens-copy separation. Auto-

197

mated equipment is designed to work within the latitude available in the lens fitted and the reprographer does not therefore normally meet depth of field problems in his work. The situation is very different in contact printing and the factors affecting sharpness of the reproduction should be his constant concern.

Requirements for contact printing

There are two reasons why perfect contact may not be obtained. First, much printing is done directly from drawn or typed transparencies, and in order to keep the copy correctly reading from left to right the copying is done with the pencil line, etc., uppermost. This means that the thickness of the tracing paper is *between* the actual lines and the photosensitive material. When printing on to an intermediate material the tracing is best turned face down, and the intermediate or sub-master also used face down when making final prints. In consequence there is a reasonable chance for good contact to be obtained, and the final reproduction prints show a better quality for this reason.

Some classes of reproduction work require the highest attainable quality and face-to-face contact is then imperative. But it is not sufficient to rely solely on perfect contact because of the inevitable presence of dust and other causes such as burred edges on cut photographic film. Even a stripping film, which transfers its layer (one half of one thousandth of an inch thick) to a new support sheet, can cause contact failure if its edge has been given a 'lift' by careless knife work.

Fortunately the imperfections in contact conditions can be largely offset by precautions in the lighting conditions, as well as aided by the mechanics of the exposing arrangements. In the early years of plan reproduction and the blueprint process, the exposure operation was carried out with sunlight in a 'sun frame'. Tracing and ferroprussiate paper were sandwiched together behind the glass front of the frame by a spring-loaded back. Contact may have been quite unsatisfactory by some standards, but the reproduction was good. This is because the sun's radiation, coming from enormous distance, consists of rays which are more or less parallel. It is for this reason that objects in direct sunlight cast a sharp shadow. On the other hand if we examine such a shadow carefully it will be found to have a slight fringe or edge which is 'half-shadow'; in this fringe area the object is not shielding the whole of the sun's emission from the surface on to which the shadow falls.

Thus we can recognize two factors which have a bearing on the ability to make a sharp reproduction by contact printing. These are the *size* of the radiation source, coupled with its *distance*. A convenient way of describing this is to define the *angle* at which any one point in the printing plane receives radiation from the source. Sharper reproduction in contact printing is obtained when so-called point sources are used, such as a compact tungsten filament lamp or a carbon arc lamp. However the radiation from even these lamps does not necessarily provide the condition of nearly-parallel rays unless they are used at a sufficient distance. The term 'point light source' is bound up with the theoretical concept of parallel light rays when a radiating *point* is at an infinite *distance*. The phrase has come into everyday use with the preferred alternative of *compact* source. Compact sources at a suitable distance give conditions in which the angle of incident light is small enough for the purpose of casting a sharp shadow; they also create a problem with regard to non-uniform intensity of radiation falling on to a flat surface.

Effect of inverse square law

This can be understood by use of the 'inverse square law', which tells us that if we have two positions at different distances from a source, the intensity at each position is in the ratio of *squares* of their distances from the source. This 'law' is a simple expression following the idea of energy radiating from a source in all directions; to catch the whole of this radiation would require a sphere, and as we go further from the source a correspondingly bigger sphere would be necessary. Thus the same amount of energy is being spread out over the surface of increasingly larger spheres. Since the total energy emitted remains steady the intensity or fraction of that energy falling at any point depends on the *surface area* of the sphere which includes that point. In, say, doubling the distance from the source, the energy concentration falls to one quarter, because the surface area of the sphere containing the new position is four times the surface area of that for the first position.

This can have quite a marked effect on the uniformity of illumination of a large frame by a carbon arc lamp. For instance, a 30 × 40in. frame might have the lamp at a distance of 50in. (this being the diagonal of the frame, which is the recommended minimum). Then the distance of the lamp from the corners of the frame is 56in., and the intensity of the illumination at the corners only 80 per cent of that at the centre. If the size of the work decreases at constant lamp distance, the effect of the inverse square law is progressively less important. For instance, at the 50in. distance an area 3 x 4in. would for all practical purposes receive uniform illumination. So that while a point- or compact-source may overcome poor contact by ensuring a sharp shadow being produced, it puts a limitation on the size of the work relative to the uniformity of illumination over the area of the work. By and large, point sources can give precise conditions only on relatively small-size work. A special condition exists in the case of printing-down on to whirler-coated sheets because the whirled coating is thickest in the centre, i.e. where the radiation intensity for a compact source is highest.

We can now see that in making contact prints there is bound to be some degree of conflict between the various 'ideals', and some element of compromise. Thus if output is the first consideration, some sacrifice in quality might be tolerated; while the utmost in quality will not be achieved at a high output. Most practical photocopying lies somewhere between the extremes of quality versus quantity. As the greatest difficulty is to achieve actual contact, there is always a tendency to use those sources of radiation which can help overcome contact loss. The nearest approximation to point sources of practical use are the carbon arc lamp and the tungsten filament lamp. When working distance relative to work area is sufficient to ensure uniformity, with silver-sensitized materials and tungsten lighting, exposure times can still be reasonably short. But with diazo and dichromate, etc., systems, exposure times can be prolonged to unacceptability. The development of plan reproduction during the 20th century is to some extent a record of how this particular problem has been tackled.

Cylindrical copiers using arc lamps

Initially the only alternative to sunlight was the carbon arc lamp; but a lamp set up to operate with a large frame does not give operational ease, apart from the lengthy exposure time. Bringing the lamp nearer to shorten the time results in uneven illumination. The answer found was the wall-mounted cylindrical copier (1896), consisting of two halves of a glass cylinder arranged vertically with a carbon arc lamp at the nominal centre line or axis of the cylinder. A canvas sheet was arranged on the outside of the cylinder with a clamping device. A drawing and print-paper, placed on the cylinder and clamped down by the canvas sheet, achieved better contact than in a flat frame because the tension applied to the flexible canvas with the clamp closed gave a compression force over the whole surface of the glass bend. The improved contact tolerated the closer lamp, while exposure around the cylinder was uniform because each part of the work was at an equal distance from the arc lamp on the axis. The lamp had to move down the length of the cylinder to equalize exposure along its axis; and by giving one, two or more passes of the lamp and controlling the rate of the lamp's descent, an adequate degree of exposure control existed.

Throughput was limited by loading and unloading, but the cylindrical copier achieved independence of sunlight. A continuous method was devised in 1920 by turning a glass half-cylinder on to its side (axis horizontal) and replacing the canvas sheet with continuous, driven belts. The glass had to be reduced from a half-cylinder to a quarter-cylinder or less, in order to bring the lamp into the correct position, carbon arc lamps still being the only suitable artificial source available. Exposure was controlled by regulating the speed of the belts. Contact might not be as good because the arc of the glass bend was small and the tension in the belts (and therefore the compression force applied to the work) was not always as great as in the vertical cylinder copier.

The arc lamp still had to move along the length of the glass bend, thought not now on the axis as was necessary in the static situation. Movement of the work around the glass bend served to equalize the exposure because the whole area of the print passed through the various zones of different radiation intensity; by the time the print emerged from the equipment, all parts had received a similar total radiation per unit area.

Use of mercury vapour lamps

One of the unattractive features of the enclosed arc lamp is the white deposit which accumulates on the inside surface of the glass, because this deposit has a high opacity to the radiations effective in photocopying. It is necessary to clean the glass frequently, which, combined with the hazards of running electrical connections to a moving object, encouraged interest in any alternative system. Eventually the mercury vapour lamp in one of its early forms found application. This was the 400-watt lamp widely used during the 1930's for street lighting. A bank of these lamps could replace the travelling carbon arc provided that sufficient care was taken to adjust the position of each lamp to give overall uniform intensity. At this time it was still common practice to trace from original drawings in Indian ink, and there was less risk of loss through over-exposure. Increasing use of draughting in pencil on tracing paper reduced this working latitude, while at the same time the larger size of the lamps increased the hazard of losses due to indifferent contact.

Exposure control

Diazo-sensitized papers showed better adaptation to these conditions than did the older ferro-prussiate process. Exposure time is varied in a continuous copier by controlling the rate of travel through the equipment. The positive-working sensitizer functions as its own exposure indicator because, being a bleach-out method, it is not difficult to determine the exposure time, which stops a little short of complete bleach-out. Plan reproduction by diazo-sensitized material works to a 'background' in the print, the density of background chosen being the *minimum* required to ensure no loss of fine detail. The background in the print is part of the image, usually being the shadow of the denser parts of the tracing-paper formation. This being so, the presence of background also serves to draw immediate attention to areas of poor contact, because at these places the residual image becomes of a uniform texture in place of the paper formation pattern. This is a condition which must quickly be recognized and remedied. Used in this way, the background image of the diazo-type print is an invaluable guide to correct copying conditions and it is unnecessary in ordinary work to use 'wedge strips' or graduated exposures as in dark-room work with silver-sensitized materials.

Modern cylindrical copiers

A next major step in evolution of the modern photocopier was the availability of tubular mercury vapour lamps and of small-diameter glass cylinders of sufficient accuracy and uniformity. It then became possible to arrange for the cylinder to be 'threaded over' the lamp and to be driven by or in conjunction with the moving belts which are responsible for providing adequate contact. The cylinder-type of machine removed the disadvantage with the glass bend, that the work had to slide over the glass surface with risk of slip occurring between tracing and print-paper.

As the radius of a modern copying cylinder is about 3in. it might be thought that placing a tubular lamp within the cylinder would be a far cry from the optimum conditions for sharp printing. The conditions are not indeed as good in this respect as might be obtained with the carbon arc in the wall copier; but they are considerably better than the copier which used a bank of mercury vapour lamps. This can be seen by considering the angle of incident light in the two situations. The actual discharge of the 400W. mercury vapour lamp is about 6in. long; at a distance of 8in. the incident angle at a point on the glass bend is about 40°. The actual discharge of the tubular lamp in the modern copier is about 1in. diameter and the lamp-to-glass cylinder distance is some 3in., which gives an incident angle of more nearly 20°. Thus the modern continuous copier does not provide as near parallel printing-light as the sun, but is an improvement on the mercury-vapour lamp/glass bend arrangement. It is, however, necessary to maintain correct and uniform tension in the belts which hold drawing and print-paper in contact with the cylinder. Even so, the best reproduction is obtained when producing intermediates or sub-masters by reverse printing (drawing face down on intermediate material) and again using the intermediate face down in making the final print. For some purposes this is a counsel of perfection; and much photoprinting is done directly from the drawing in face-up position with satisfactory results, so long as the correct degree of background is left in the print. If it is desired to test the ability of such a copier to give satisfactory conditions, it will be a useful exercise to print from an overlay of, say, three photographic film positives (line subjects). While the image produced from the top and middle film will not be of the quality obtained from the lowest film, this method can give acceptable results when such a 'composite' copy is necessary.

Requirements of precision copying

The type of equipment described represents a compromise between the conflicting requirements of 'critical' lighting, adequate contact, convenience and throughput. Because of the deliberate acceptance of prints with background when using diazo-sensitized materials there is no serious risk of unsatisfactory results, while a fully bleached-out print (i.e. no background) will entail certain loss of detail.

There are, however, applications of photoreproduction methods which cannot tolerate background and in which it is essential to retain fine detail. Continuous or rotary copiers then seem attractive, but may not provide the class of result necessary for the purpose in hand. This may be the 'dot-for-dot' reproduction of a screened negative, or the preparation of an etching resist in a miniaturized electrical component. In these and similar applications, the objective is accuracy, with maximum possible density in the image parts and minimum possible density in the non-image areas. A density higher than necessary in the non-image parts can have a worse effect than low density in the image; for instance, a film positive for printing down to deepetch lithographic plate has to create a clean etching resist. A high background density may require prolonged exposure to obtain the necessary resist to protect the metal surface from the etching solution. If in so doing, unwanted image is formed where etching is to take place there will be incomplete etching of the parts which are intended to print.

To meet the requirement for precision copying, by which is meant not only a sharp and accurate image but also an absence of unnecessary background image, one has to use the printing frame and near-parallel radiation for exposure. Thus from the beginning of photoengraving, the work was done on relatively small pieces of metal. Simple spring-backed frames were sufficient to bring the glass negative and sensitized metal into reasonable contact, and exposure made to a carbon-arc lamp at some six or more feet distance. The small size of the arc source relative to its distance, and the small size of the work itself, was able to compensate sufficiently for any shortcomings in contact. Exposure times were of the order of ten or more minutes, but this was not objectionable, being only a small fraction of the overall time of preparing the finished printing block from the plain metal sheet.

Vacuum frame for improved contact

Any method of improving contact made it possible to shorten lamp-to-frame distance, with exposure times rapidly reduced in accordance with the inverse square law (halving the distance increases the intensity at the frame by four times). Apart from more robust frames and correspondingly stronger springs to apply pressure to the frame back, the great step forward in improving contact came with the vacuum frame.

Vacuum frames have come into very wide use, not only in process shops devoted to photoengraving or preparation of photolithographic plates, but in small-size maximum-convenience units available as part of any small-offset lithographic printing department. A large proportion of the vacuum frames installed are not operated to maximum advantage and the principle on which they work should be understood by the operator.

As with any printing-down frame, there is a rigid framework which holds a glass front. The back or bed of the frame is a flexible sheet of reinforced rubber, etc., with a dimpled top surface. When the frame is closed, air is extracted from the enclosed space by a vacuum pump; there has to be an effective seal round the edges of the frame to make it possible to remove air from within the frame space. The dimpled surface of the flexible base sheet ('blanket') allows air to be pulled rapidly from all parts of the frame area and avoids pockets of air being trapped between the glass front and the blanket.

Use of backing sheets

The effect of evacuating air from the frame is to create a pressure differential between the external atmosphere and the space within the frame. The maximum pressure which the atmosphere can exert is some 15lb. per square inch, so if evacuation were complete, this pressure would be applied over the whole front and back surfaces of the frame. The flexibility of the rubber blanket allows this pressure to be applied uniformly to the work in the frame. The common fault encountered in practice is that the work is often supported on a sheet of rigid material, under the mistaken impression that this will increase the pressure applied. On the contrary, the effect of a rigid backing is to hold the work between two rigid materials (glass front and backing sheet) which are *spaced apart* by the maximum thicknesses of the work itself. Thus any overlapping of two pieces of film, or taping of a negative to a masking sheet, will form spacers and the work itself may lose contact disastrously despite a high vacuum being achieved.

Although rigid backing sheets must be avoided for efficient use of vacuum frames, some backing is desirable because the dimpling of the blanket may create pressure points. The function of the backing is then to distribute the pressure away from these points and create conditions of uniformity over the area of the work. For this purpose, a material of some compressibility is desirable: pieces of smooth felt or thin sheets of certain 'foam' materials. Today much printing in vacuum frames is from film negatives or positives on to metal sheets. Printing-down from film to rigid metal arises in photoengraving and

205

it then becomes necessary to remove anything which may space the film and metal apart despite the vacuum effect. It is also becoming frequent to use vacuum frames in copying from film to film. In this case the work is very flexible and the conditions for obtaining close contact are very good; but apart from a backing which is flexible it is then very useful to have one which is black, orange or yellow in colour. This absorbs actinic radiation which penetrates the two films and which might otherwise be reflected back in a scattered form, thus reaching parts of the sensitive layer which should be 'protected' from radiation by the dark parts of the original film.

The importance of a black backing sheet in film-to-film work can easily be seen if one part of the work is backed with white paper or bright metal foil and another part with black paper or felt. With, say, diazo intermediate film materials, there will be distinct differences in exposure time to achieve a fully bleached background. But apart from this, if exposure needs to be longer than the bare minimum for nil background image, the outlines of the image over the white backing will become less sharp and image density will suffer.

Procedure for critical reproduction

Assuming the vacuum frame is used with the care indicated, and of course that image-to-surface conditions apply, we then have come as close as is possible to the conditions for perfect contact. It may no longer be necessary to use the near-parallel ray condition. Unfortunately, the vacuum frame cannot overcome the effect of dust; so that in critical work such as very fine screened subjects a distant 'point-source' might be obligatory, whereas a bold line subject might be quite satisfactory with a bank of fluorescent tubes. Conditions to suit many situations can be found between these extremes; the temptation is to choose working convenience and speed rather than technical suitability. If quality and critical reproduction are wanted they will be achieved only at a relatively low output rate.

The contact reproduction of screened photographs is probably the most critical work of this type to be undertaken. If the tonal values of the subject are to be retained, the magnitude of each dot must be correctly reproduced. Exposure conditions must avoid any tendency for spread or undercutting of the image. Usually this work is done with silver-sensitized materials, for instance a screened negative is prepared as a means of producing a screened positive.

In the negative the highlight dot is a small opening in the heaviest parts of the image while the shadow dots are small black areas. It certainly requires the condition of parallel-beam illumination to achieve accurate reproduction in such cases. Fortunately the silver-sensitized materials have a good response to tungsten filament radiation, combined with the amplification effect of development. This makes it possible to contact-print with a small bulb at a considerable distance: say, a motor-car headlamp bulb at six or more feet from the frame. The lamp would also be housed in a box painted black on the inside so as to eliminate stray light reflection, relying on direct radiation from filament to frame.

Effect of glass in photocopiers

In all methods of carrying out contact exposure it has been necessary to interpose between work and radiation source a glass bend, cylinder or plane sheet as an essential in achieving the necessary contact. This glass may well be $\frac{1}{4}$in. thick and sometimes more. The only radiation which may reach the photosensitive substances of the copy-sheet will be that which the glass transmits. Glass normally has a high transmission of visible light, but not necessarily to the radiation regions beyond the visible spectrum. The ability of any material to transmit radiation is best expressed, like the properties of radiation sources or of sensitive materials, as a spectral diagram involving a base scale of wavelength.

Glasses are a group of chemical substances which can have widely different constitutions and correspondingly different behaviour in respect of radiation transmission. The glass normally encountered for window glazing has a poor transmission beyond the visible violet. Specially-compounded types of glass can have a higher transmission beyond the violet and are sometimes claimed for that reason to be 'healthier' for domestic use. The glass used for photocopier cylinders is certainly of that type, and requires to be 'worked' at higher temperatures than normal glass. This conflicts with the requirement in photocopier cylinders for accuracy and trueness as well as freedom from flaws such as air-bubbles or dirt inclusions. Vacuum frames tend to be fitted with plate glass, which may not have as high a transmission of actinic radiation, but is produced with accurate plane parallel surfaces.

With the exception of the open carbon arc, light 'sources' have to be contained within an envelope of transparent material, usually glass. The enclosed carbon arc has glass about $\frac{1}{8}$in. thickness; tungsten filament and discharge lamps have much thinner-walled glass envelopes which nevertheless fail to transmit as much of the radiation as might be desirable. Quartz can be used to retain a high transmission into the ultra-violet regions, but contributes a high cost and a certain amount of danger through its greater fragility. The curve on this page is of the radiation passed by 2mm thick Pyrex glass.

Filtering out unwanted wavelengths

Thus the glass of frame, cylinder or lamp is an unavoidable item in the photoreproduction situation, and removes or 'filters off' some of the available actinic radiation. Additionally, it may be desirable to

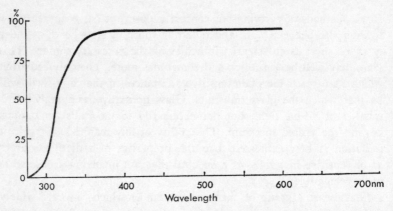

Spectral transmission of 2mm thick neutral Pyrex glass. A 10mm thickness of water has 95 per cent transmission throughout this wavelength region.

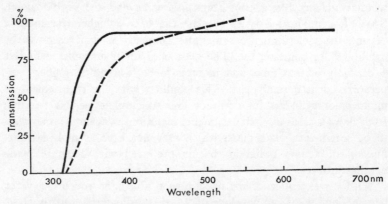

Spectral transmission of plate glass 3.05mm thick (solid line) and of polyester film 0.004in. (1.016mm) thick (broken line).

208

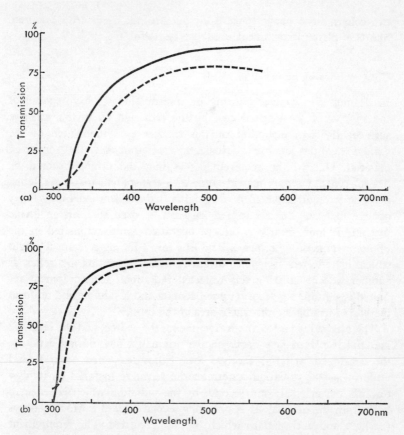

a, Spectral transmission of cellulose acetate 0.143in. thick (broken line) and of cellulose nitrate 0.147in. thick (solid line). *b*, Spectral transmission of polystyrene (broken line) and polymethyl methacrylate (solid line).

restrict deliberately the wavelength regions which are allowed to reach the photosensitive surface. A material for this purpose is referred to as a *filter* because it is chosen to transmit certain radiations but not others. The most obvious use of filters is in the camera room, when special effects are wanted and when coloured subjects are being reproduced. But it also happens that a filter may be beneficial in the contact-frame. For example, a radiation-source may be rich in wavelengths to which the copying material is sensitive, but which the original image transmits to a certain extent. By placing a suitable filter in the frame, these wavelengths can be absorbed and turn an 'impossible' situation into an acceptable one. Filters are used in a two-colour diazo paper (page 155) but are not necessarily coloured. Sheets of plain plastic, uncoloured, can be useful.

Contact printing by reflex methods

In dealing with contact printing by transmission we have spoken of the *shadow* of the original cast by the radiation on to the sensitive surface; the reproducibility of this shadow is compounded of the various qualities of the original, the radiation and the sensitized material. One of the requirements is that the original should be drawn, typed, etc., on one side only of a transparent material. Reflex printing is required when the original is two-sided, usually on a fairly opaque material. In this case a lens can be used to form an image but this is more nearly a camera operation (semi-automated in the photostat type of equipment). The lens forms its image from the light which is reflected from the original; contact printing by reflex is another way of utilizing this reflected radiation. The equipment required is a frame for securing good contact and a suitable and uniform method of irradiating the whole area of the frame.

The original is laid in the frame, facing the source, and the sensitive material is placed *face down* on the original. Thus, during exposure the sensitized material receives two exposures: first, overall and uniform, as the radiation penetrates the sensitive material on its way towards the original; and second, by the radiation which is reflected back from the original. It is only this second exposure which conveys the light and dark pattern which is to be recorded. The requirement is for the photo-system to 'ignore', or not respond to, the first exposure and to function only on the basis of the second or reflex exposure.

This is not a difficult position to achieve using silver-sensitized materials. The ability to differentiate between the two levels of exposure may be increased by filters – that is, by restricting the radiation involved to a fairly narrow wavelength band. But the main factor is the characteristic of the photographic emulsion and its chemical process of development. Reflex printing by the diffusion-transfer technique has been very popular.

Reflex printing with diazo materials

Attempts have been made to use diazo-sensitizers in reflex copying, and have succeeded up to a point but not to the level required for a working process. This is because the diazo process by dye coupling has too low an amplification factor to succeed in producing an acceptable record of the available pattern of light and dark areas. Hence the popular diazo materials are restricted to transmission printing and to projection printing with specialized equipment.

It has been made possible to reflex-print with diazo layers by a 'screening' system invented by the Dutch firm of Van der Grinten. This required that the diazo compound be coated on film and that the exposure be made under a very fine screen. In the early form, the screen was formed on the lower surface of the glass of the frame; it can be visualized as a continuous deposit of metallic silver, with numerous minute openings.

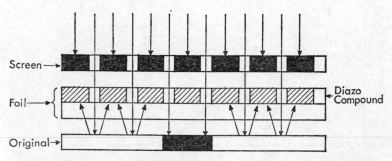

Van der Grinten system of reflex exposure with diazo sensitized foil and screen.

During the first exposure, radiation only reached the diazo layer in the areas immediately below the screen openings. When these parts were fully bleached, the radiation was either absorbed by the dark parts of the original, or reflected by the light parts. Where the radiation was reflected back, exposure began of the diazo compound hitherto protected by the solid parts of the screen and had to continue until these parts had also been completely bleached. Thus on dye-coupling (one-component system), no dye should form over the light parts, and maximum dye over the dark parts. However, the whole image formed was necessarily broken up by the first exposure through the screen openings. Being on a film base, the resultant image formed an excellent intermediate for making normal diazo paper copies, in which case it was coupled to form a sepia dye. For single copies, a film with a latex backing was used, which after exposure but before 'development' was laid down on to latex-surfaced white paper. Coupling then formed a black dye mixture.

In later versions of the Van der Grinten reflex-diazo process, the screen was formed on the top surface of the diazo-sensitized film and comprised a mixture such as carbon black and asphalt. After exposure, this screen was removed from the print. It will be clear that the maximum density obtainable is reduced by the areas of the screen openings, which amounts to between 10 and 20 per cent of the total. The best results in terms of density and contrast are with screens of between 125 and 200 openings per centimetre; 15,000 openings per square centimetre is frequently quoted.

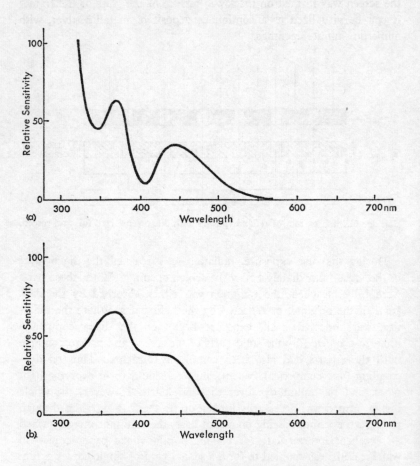

Change in effective spectral sensitivity of ammonium dichromate/gum arabic system at 25°C (*a*) and 35°c (*b*). M. Alexander, *Influence of light sources on printing-down in offset-printing* (in German), Deutsche Gesellschaft fur Forschung im Graphischen Gewerbe, 17/5, pp. 18–20.

Heating effects in contact printing

All sources of radiation useful in photoprinting have some proportion of their output in wavelength regions to which the sensitive material does not respond. This energy is absorbed by the exposure equipment, resulting in appreciable rises in temperature. In the case of dichromate/colloid systems which have a 'dark reaction' this may be important. The higher temperature may well impose a different spectral response through acceleration of the process of the dark reaction. This is shown graphically by the diagrams of M. Alexander, representing a rise of only 10°C. in the temperature of the frame. So far as possible it is best to choose the radiation source which has the highest proportion of active radiation to inactive. Unfortunately the most efficient sources from this point of view are the carbon arc lamp with its variability and the fluorescent tube with its large angle of incident radiation (see page 199). The answer to the heating problem is to cool the printing equipment with air – not difficult in a rotary copier provided that local cooling of the discharge lamp is avoided. Vacuum frames may also need to be cooled, especially if printing from a xenon lamp.

CHAPTER 14

PREPARATION OF ORIGINALS

The most commonly-met originals today are pencil drawings on tracing paper, and typewritten documents on a variety of translucent papers. However, this is by no means the whole extent of the variety of originals and additions are continually being taken into everyday use. As pencil drawings and typewritten work have certain basic similarities we will consider them together before listing other types.

Initially a tracing was a copy produced on a more or less translucent paper or prepared linen sheet, laid over a drawing. This method of obtaining a copy dates from the 15th century. More recently, tracings have been used to produce clean copies, preferably in Indian ink, of a working drawing with a view to photocopying. The method is no longer in widespread use, but tracing paper and tracing cloth are freely available and still so-called. The modern method is to *draw* on the tracing material and use this for making the photocopies.

Thus tracing papers or similar (e.g. detail paper, or 'prepared' papers made translucent by impregnation) have become the most commonly used materials for original drawing or typing purposes. In the main, pencil is used on tracing paper and the drawing is expected to reproduce well in photocopying. The preparation of an intermediate (see page 158) from the drawing might be said to have replaced the former practice of tracing in ink. Provided that face-to-face contact is used and correct exposure given, the reproduction quality of an intermediate can be higher than that of the original drawing. This has given rise to the concept of intensification or strengthening of pencil work by means of an intermediate copy.

Typewritten originals

In addition to pencil or ink drawings it has become increasingly common practice to make diazo copies of typewritten matter. The normal 'black' ribbon of the typewriter is of little use unaided, because it has a high transmission in the blue and violet regions to which a

214

diazo compound may be sensitive. Ribbons for this purpose are of a brownish-black colour and give a high colour deposit under the pressure of the typewriter character. If such 'reproduction' ribbons are not available, a usable result can be achieved by backing the impression with a yellow carbon paper (i.e. arranged behind the sheet to be typed, carbon-side uppermost, so that the carbon is transferred to the back of the typed sheet). Disadvantages of the yellow-backed work are that there may be transfer of the carbon where it is not wanted, especially if it creases in the machine to give 'treeing'; and the possibility of the carbon being smeared in handling or in the copier itself. Properly handled, the reproduction ribbon is preferable, giving the better result and cleaner work.

Typing may be on tracing paper or on the bond-type papers now available. The latter are of a white appearance close to the colour of normal stationery, not fragile as tracing paper tends to be when dry, and with a sufficiently high ultra-violet transparency for reasonably rapid print-off.

When an organization equips all its typewriters with reproduction ribbons and adopts suitable paper for letter-headings and internal use there is little need for more than one normal carbon copy to be taken. Any other copies required can be photocopies all of equally good legibility, instead of in decreasing quality to the sixth or more carbon copy. If the original typing can be retained in the file and photocopies distributed, it is always possible to produce further copies of equal quality whenever required during the life of the document.

Pencil and typewritten work all suffers from one defect in that the line or character is not 'solid' or continuous. The pencil line is produced by deposition of the graphite composition on the high spots of the paper but not into the low spots. The line is thus broken into small dark and light areas and has undefined edges. The typewritten character is similarly impressed on the higher parts of the paper surface and may be further broken by the fabric of the ribbon used.

The reproducibility of the pencil line is increased if it is drawn on a more level surface, provided the surface has sufficient 'tooth'. It has been possible to provide such draughting surfaces on film, which itself offers a more uniform thickness as a basis for the top coating. Such draughting films, using polyester base, are now widely accepted where high quality in both draughting and reproduction is required. The typewriter can go further by the adoption of 'film ribbons'; these are of thin polyester film carrying a pigmented layer not unlike that of carbon paper. Under the impact of the typewriter character, the tendency is for the whole of this layer, in the shape of the character, to transfer to the typed sheet. Thus no fabric pattern can arise and a well-filled impression of sharp outline is produced.

Conversion methods

An even better result can be achieved from the film ribbon by typing on a paper with a really smooth surface. This, however, entails a coated paper such as the baryta paper of the photographic process. However, while the quality of the impression is thus improved, the paper is usually too opaque to allow contact-printing, and the work has to be photographed by camera to carry it on to any further reproduction stage. For ordinary purposes the time and cost involved in such a procedure would be prohibitive, but in certain situations it is attractive. 'Information' printing – instruction manuals, etc. – is increasingly done by such methods, making for quick availability and lower cost than conventional letterpress. This type of operation is conveniently grouped under the heading of 'conversion' methods, the term referring to any method whereby a type impression is converted to a photographic film or transparency suitable for the making of a lithographic plate.

Some of the conversion methods used are purely mechanical in operation but several are reprographic in character and form an expanding area of photoreproduction. Phototypesetting if generally available would make the conversion processes unnecessary, but 90 per cent or more of typesetting must still be carried out by the assembly of metal type into words and their arrangement into lines, paragraphs and pages. In addition to the requirement of converting set type to a lithographic printing surface, a similar problem arises when existing engraved blocks – line or screened – are to be incorporated. It may even be that set type and engraved blocks require to be used for the preparation of a single, new, relief printing surface produced by powderless etching; in this way curved or wrap-around printing formes can be prepared by reprographic methods, from relief surfaces in the flat.

Influence of original or reproduction method

The techniques of the photocopying intermediate or sub-master allow the combination of items, selecting or rejecting areas of the original work, and also assist in the introduction of a second colour or shaded areas, etc., into the finished job. Thus reprographic methods bring to the general field of graphic reproduction, techniques in which photo operations replace or supplement manual operations.

Before describing some of these in more detail, it is appropriate to review the types of original so far mentioned and the particular conditions imposed by the original on the reproduction method employed:

1. Normal drawings in pencil, on tracing papers or prepared draughting films
2. Typewritten matter on various papers, with normal fabric ribbons and reproduction ribbons
3. Typewritten matter on smooth papers with 'film' ribbons.

Because of their base transparency, types 1 and 2 would normally be reproduced by transmission contact printing using one of the diazo (positive-working) materials. As type 3 usually uses more opaque paper, reproduction normally requires the camera giving a negative, or silver reflex material which can be of the auto-reversal (positive-working) type.

Straight contact prints of types 1 and 2 would be made by exposing with the work uppermost, retaining correct left-to-right reading in the copy, but requiring that the base material lies between the drawn or typed work and the sensitized surface of the copy material. The result can be tolerable only to the extent that exposure conditions approximate to parallel radiation. Notice, however, that type 3, if photographed in the camera, should not suffer in the same way; the image is focused by the camera lens directly on to the surface of the sensitized material or, if reflexed, the original typed character is necessarily in contact with the sensitized surface. Thus the *normal* reproduction of type 3 originals is better able to give crisp, sharp results than is possible from types 1 or 2, so that the superior quality of the film-ribbon character on smooth paper is fully retained.

The quality of the normal reproduction of types 1 and 2 thus depends upon how far the exposing radiation approaches the parallel condition, and the extent to which the base through which the line must cast its shadow is non-scattering and non-diffusing. Moreover, the drawn pencil line is not continuous or of uniform width and the typed character is not crisp and may be discontinuous. In both cases therefore the original is broken and the reproduction can be brought about only by the imperfect shadow cast by the broken line. Assuming a positive-working material for reproduction, the danger is that the shadow will protect insufficient of the photosensitive material. Hence

217

the invariable convention with these types of leaving background in the print; the copy then depends upon the visual difference between line-areas which have been partially bleached and background which has been more nearly fully bleached. Reproductions from pencil or typed originals cannot utilize the full contrast of the reproduction material, or in other words cannot retain the maximum density available at the minimum density possible. The yellow carbon-backed typewritten master has an advantage because the yellow part of the impression is adjacent to the sensitized surface but it is not capable of protecting fully from the actinic radiation. The intermediate technique improves the position with any type of original.

Assessment of materials

We may note in passing that because the reproduction of pencil tracings cannot utilize the full contrast of the diazo material, the manufacturer formulates for this situation and most normal products are to be assessed or compared with respect to actual pencil-line reproduction. It is impossible to select diazo copy papers on the basis of maximum available density on fully bleached background. Neither the colour nor density of the solid accurately reflect the visual impact of the line in print. This should not be difficult to understand in view of what has been learnt of the visual process (see pages 15, 16).

Similarly, the assessment of intermediate materials must be on the basis of the opacity or absorption of actinic radiation by the line produced from original pencil lines, and not on the maximum density available (though this is of importance in a later context, see page 223). To compare two intermediate materials requires that both should be exposed under the original to matching background density – probably at different copier speeds or exposure times – and then printed down side by side on to single sheets of the final print paper. A series of these final prints at varying background density is required. In this way errors due to lamp fluctuations or the like are eliminated. The final results must then be related to any differences in print-on speed when making the intermediates.

Originals for high-quality contact printing

The question arises whether any working method may more nearly utilize the full potential of the reproduction process when contact printing. This must involve an original which is superior to types 1, 2 or 3 as described, with respect to the sharp outline of the character or line, complete and uniform 'fill' within the outline, and adequate density for the intended process. In this case 'process' means the combination of sensitized material with radiation source and any limitations of the method, such as glass used in frame.

The most obvious original which might fulfil these conditions is a well-made photographic film, positive or negative as the case may be. Although we still refer to this as an original, it is of course itself a photoreproduction and we are now in the area in which successive reproduction stages are used, but not necessarily the same type of reproduction system at each stage. There are other ways of obtaining originals of the stated character than by using silver-sensitized layers, not always as near to perfection but eminently acceptable and avoiding the complications of silver-material processing. In all these cases we now apply a new criterion of quality, namely that of the extent to which the original makes possible reproduction with the highest possible image density retained on minimum possible background density. The minimum possible background is that of a fully bleached-out positive-working product, or the unexposed parts of a negative-working product; in either case, full normal processing must be used before judging the minimum density attainable.

Originals which may fulfil these requirements comprise pigment or metallic image areas on high-transparency base materials. Of the pigments, carbon black is the most frequently met and of the metals, silver. We can add to the list of originals:

4. Well-made drawings in Indian ink, etc.
5. Tape drawings, as used for printed circuitry
6. Pigmented impressions, as of printing ink, with or without powder reinforcement
7. Masks, etched from thin sheet metal or by deposition on glass
8. Photographic products, especially thin-film layers of lith type
9. Pigmented scribe or peel films.

These and others may demand different handling by the reprographer from that encountered in normal plan reproduction. Probably they are intended to operate within defined limits which can be met only by precision printing as already defined.

Production of pigmented impressions

Of the above types of original, probably only 6 requires any explanation. These are usually originals prepared by letterpress printing for purposes of further reproduction. Hence the term 'repro' as used by printers, meaning that the printed sheet is destined to be an original in one of several possible ways. The operation of printing for this purpose is carried out with care, to avoid any smudging, etc., which would detract from the end-result. Every effort is made to start with a good-quality impression from new type and to retain the quality in subsequent stages. Because only a few impressions may be taken from the type or blocks involved, there is a tendency to speak of them as 'pulls' in the sense of proof pulls; hence 'repro pull' or 'repro proof'. But even though a proofing press be used for making the repro pull, it will be used with skill to obtain as clean and as near perfect an impression as possible.

In the simplest case a completed page or pages of type would be 'pulled' for repro on, say, baryta-coated paper. This could be photographed in the camera to give a negative for platemaking, possibly with enlargement or reduction in size. Alternatively, a contact print on silver-sensitized material could be made for same-size reproduction. In contacting, the ink impression is of course face-to-face with the sensitized surface. Another use of the repro pull from type on to baryta paper might be to form a paste-up with line or screened illustrations on normal bromide paper, the whole then being photographed to give a single composite negative.

Repro pulls on H2 film

Although these methods of work are reliable there is always a desire to avoid photographic (darkroom) operations if possible. It would obviously be attractive to take a repro pull on film and use it directly for making the lithographic plate. The normal film surface does not, however, approach the surface of paper with respect to the ease of taking a sharp, clean impression. Even if the ink film from each line or character is uniform, it will not have the density required for platemaking. It is possible to arrange that the film receives two impressions of the type: one on the back by offset, and one on the front direct from the type. A more usual way is to take the impression and reinforce the ink by dusting over with a black powder.

The widely known H2 film is $1\frac{1}{2}$ thousandths of an inch thick, and carries on the top side a thin resilient coating. This coating allows some compression when the sheet is printed, avoiding excessive ink squash, and recovers itself as the printing pressure is removed. Thus the impression remains sharp, though the dusting tends to produce some spread and to fill in to some extent small openings in a black area. The intention with such a printed sheet is that the reinforced ink image will be capable of making a lithographic plate – frequently by the deep-etch process which uses a positive. However, to produce a plate for offset lithography requires an image on the plate which reads correctly from left to right; therefore the plate has to be made from the printed film by exposing through the back of the film. Even though critical conditions are used, some further loss of quality seems inevitable through this $1\frac{1}{2}$ thousandths of an inch separation of the ink/powder image from the sensitized plate surface. One way of retaining image-to-surface contact would be to reverse the image laterally by an intermediate step, analogous to the use of intermediates in plan reproduction. For this one could use an auto-reversal (auto-positive) silver film for transmission printing – as with the typed baryta paper which was reflex-printed; or a normal silver film to give a negative; or a diazo film to give a positive.

Comparison of repro pull methods

Note that the baryta pulls plus photographic stage (camera or contact) necessarily avoids the problem of contact loss. The film pull used alone introduces out-of-contact platemaking, but film pull plus an intermediate stage restores the contact condition. Obviously baryta paper plus film is cheaper than film plus film; but whereas normal baryta demands a silver-sensitized material on account of its opacity, the repro pull on film allows contacting to a diazo intermediate. In terms of material costs alone, the alternatives are:

For single sheets, 24 x 20in. or equivalent area

		new pence
1.	Baryta paper	2.5
	Silver film for contact or camera negative	80.0
	Total	82.5
2.	Repro film alone	40.0
3.	Repro film	40.0
	Diazo film for contact	32.5
	Total	72.5

221

Additionally, processing the silver film may cost about 50 new pence. Thus the demand for quality to be obtained by method 1 must be high enough to sustain the higher cost: some four times that of film alone if printing to plate through the back of the film, and about double the cost if a film plus diazo intermediate is used to retain face-to-face contact at each stage.

It is a natural thought to enquire if a baryta-paper repro pull, with its inherent good quality of impression, could be combined with a diazo-contact intermediate. If so, quality and economy can be combined. The ordinary baryta paper has too little actinic transmission, but a special type of baryta-coated paper is available in which impression quality is fully retained, and actinic transmission much improved. By selecting the type of radiation and diazo sensitizer, intermediate film copy can be made from such baryta pulls with exposures of three to five minutes instead of 15 to 20. As the diazo product has virtually no processing time, and no washing or drying, the preparation of film intermediates by contact from these baryta-paper repro pulls becomes an attractive proposition. This gives a fourth alternative with material costs:

		new pence
4.	Special baryta paper	5.0
	Diazo film for contact	32.5
		Total 37.5

In this case, the quality will be as for 1, but at a cost no more than for film used alone.

No doubt a higher density could be obtained from a silver copy as example 1, but the real comparison is with the H2 film result which is of sufficient density for platemaking, as is the diazo result if correct material is chosen.

Characteristics of Scotchprint material

For completeness, mention should also be made of the repro film material Scotchprint. This is a polyester film with a white coating. The coating has excellent printability and will sustain impression quality even if heavy printing pressures are used. After impression, it is used for contacting to silver film for a negative, or to auto-reversal

222

film for a positive; or photographed in the camera to give a negative. Unfortunately its ultra-violet transmission is very low, and contacting to diazo film intermediates though feasible is a process for only occasional use.

Retaining contrast in originals

Originals produced on photographic film, especially lith-type film, should consist of an image of uniformly high density (preferably the maximum of which the film is capable) and non-image areas of uniformly minimum density. By many reprographic standards, copying from such an original is simplicity itself, and has much greater latitude – room for error – than copying from pencil drawings on indifferent tracing paper. But the aim of further reproduction from such an original is to retain *exactly* the sizes, shapes and density of the image areas without increasing the non-image density. This may require considerable adjustment in outlook for the average reprographer, but it is becoming increasingly important to make this adjustment owing to the expansion of the common ground between photoprinting and printing.

The image of a silver-sensitized film comprises metallic silver particles distributed in a thin layer of gelatin. The silver particles vary in size and form according to the emulsion and the method of development. It is possible for two pieces of film, processed differently and equal in density at the visible wavelength of measurement, to be very different in their densities at about 400 nm. Frequently the absorption of actinic radiation by a silver image is *higher* than its absorption of visible light; thus in viewing a silver film positive, its capacity for subsequent successful photoreproduction may be greater than would seem from visual examination or normal densitometric measurement, which relies upon a visual calibration of the densitometer (see page 224).

Should a high-contrast silver film positive be used for making a diazo film intermediate we must be careful to see that the diazo image has, in turn, the highest density of which the film is capable; and the non-image correspondingly the lowest possible density. It should not be difficult to ensure this on account of the UV absorption of the silver being greater than its visual density. It may also be possible to use it to improve the diazo copy with respect to the non-image areas; if the processing of the silver film has given localized areas of high fog level, they may be removed in the diazo image by prolonga-

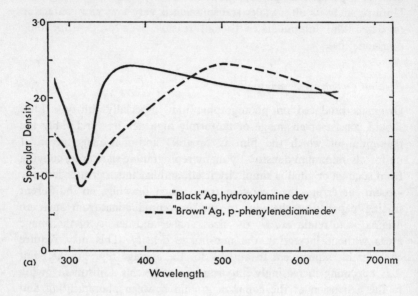

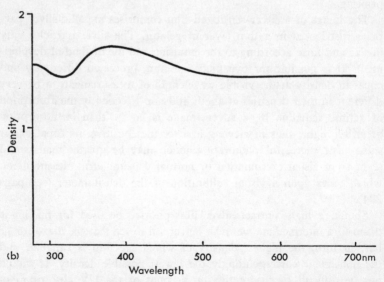

Spectral absorption of silver images: a, influence of development conditions on spectral absorption of silver image (high contrast. motion-picture positive film). b, microfilm negative emulsion.

tion of the exposure time, without detriment to the maximum diazo image density. This makes possible a number of interesting techniques of silver in conjunction with diazo in the preparation of printing surfaces and for photomechanical purposes. At the same time, it becomes necessary to watch the type of subsequent sensitizer and the nature of the radiation source which will be used.

In practice it can be quite satisfactory to print down from a positive diazo film intermediate on to, say, deep-etch litho plate (dichromate sensitizer, negative-working) using carbon arc lamp. Greater working latitude would be obtained in printing down from diazo film intermediate to a presensitized litho plate (diazo sensitizer, negative- or positive-working) and carbon arc or xenon lamp can be satisfactory. Even greater working latitude would be given by using a diazo presensitized plate and fluorescent tubes for the printing-down operation. These possibilities are to be understood by making use of the relevant spectral curves: radiation source, diazo image spectral absorption, and metal sensitizer spectral sensitivity.

Avoiding film-edge reproduction

One of the more acute problems which has arisen with the use of phototypesetting is that of 'film edges'. When a sheet of photoset film is produced it has to be cut up and reassembled in the required lay-out of paragraphs, pages or columns, etc. For maximum economy it may seem desirable to use the assembly of film to print down on to the lithographic plate, an operation which traditionally employs a carbon arc lamp or, nowadays, mercury or xenon discharge lamps.

The compact light sources are valuable in overcoming losses of contact which may occur even in the vacuum frame; but they result in the four edges of each piece of film being reproduced on the plate. The effect is purely optical, because there is nothing at or in the film edge to reproduce. It is caused by total internal reflection at the edge, which is more pronounced when the radiation is of more nearly parallel character (just as the image of the vesicular diazo-sensitized film is best suited to projection purposes and has increased effective density when the aperture of the projection lens is reduced). The result of this internal reflection process at the edge is to divert rays which strike the cut surface at less than a given angle – in much the same way that a stone can be 'skipped' off the surface of water. Thus the areas immediately beneath the cut film edge are deprived of radiation which arrived at the top surface of the film. With a positive-working system such as the diazo process, when printing from a well-made silver image it is perfectly feasible to prolong the exposure beyond the minimum for clear background so that the 'deprived' areas have the opportunity to reach the condition of total diazo decomposition. Then the optical shadow formed by the cut film edge is not carried into the new image. Thus, a reprographic process can be of great benefit in photomechanical situations by avoiding the necessity for edge lines to be removed by tedious handwork – either by painting-out as on deep-etch process, or by chemical deletions on others.

As film-edge reproduction is encouraged by the near-parallel radiation chosen for maximum accuracy, the problem can be diminished by departure from this condition. Thus a bank of fluorescent tubes is from this point of view preferable to a compact source; much of the radiation reaching the film edge then exceeds the angle of incidence below which the total reflection phenomena exists. The drawback, however, is that any failure in contact, such as is caused by dust, results in an unsharp shadow or 'undercutting' of the true image. Under reasonably clean conditions this may not cause difficulties in reproducing from text; but screened half-tones suffer in that the dots near the dust speck are not as crisp and clean as required for correct tone values.

A practical middle course is to use the compact source, but to place a diffusing screen between the work and the source. The screen may be inside the frame if required for the whole exposure, or on the outside if needed for only part of the exposure time. Thus varying degrees of departure from parallel-radiation conditions are possible, and a few trials will establish a new skill in techniques for this type of photoreproduction. The draughting films of the drawing-office make good diffusers for this purpose.

Multi-colour printing

Once the principle of working from phototypeset silver film via diazo-sensitized film intermediates has been accepted, other time-saving techniques become possible. For example, it may be required to print a text in two or more colours. Obviously all the different coloured impressions must come in the right places on the finished article. An excellent way of ensuring this to make a number of diazo-film intermediates of the total film assembly, but after exposure and before processing through the ammonia machine, to give each a second exposure under a mask. The mask is made to protect during the second exposure those parts which are required to print in one particular colour. As all the intermediates have been made from the one original film assembly, the various partial images must remain in correct relationship to each other. A punched-register system ensures their coming together correctly on the plate. To make the masks, peel-coated films are excellent: the parts removed are easily placed on a clear film as one mask, while the parts remaining form the other. Here again the materials and methods of reprography provide valuable techniques in photomechanical situations.

The diazo processes can show further flexibility in giving a dye image which has very *low* actinic density, i.e. has a relatively high transmission of radiation in the wavelength regions of 450nm and lower. For some time the method of making a 'blue key' as a registration guide has been to use a dichromate/colloid layer on film, and dye the image remaining after development. In this way a key-sheet might be made of, say, the black-printing parts of a two-colour diagram; the parts to be printed in the second colour can then be attached to the blue key sheet with complete accuracy. When making the plate for the second colour, exposure is prolonged to print out the blue key image. Diazo-sensitized film can be made which produces a blue key image by simple exposure and ammonia processing, i.e. without the wet-processing of a dichromate-colloid system or the messy dyeing bath.

These examples show how suitable choice of materials allows a high-density image to be converted to special-purpose images in an alternative system. Without doubt there is an increasing area for applications of this kind, as the fields of printed circuitry, chemical milling (photo-fabrication) and specialized 'drawing' expand alongside the more conventional fields of photoengraving, photolithography, photogravure and silk-screen printing.

Methods of photo-drawing

There is a continuous cross-fertilization of ideas and techniques between photoprinting and photomechanical situations. One such example is the wider adoption of 'photo-drawing' in the drawing-office. Photo-drawing requires a photograph of an actual article, which might be the prototype of an electrical system or a scale model of a chemical plant. The intention is to use these photographs instead of making conventional plan and elevation drawings but to incorporate the photographic 'information' into a photoprinting original, so that photocopies may be distributed and handled in the normal way. The photographs used must be skilfully made with respect to camera angle, perspective and depth of field. As the final copying will probably be on diazo-sensitized papers, the photograph has to be a positive transparency either of low contrast to match the high contrast of the diazo reproduction or screened to utilize this contrast.

There are several ways of using the photograph, and all are reminiscent of techniques used in the printing industry. For example, one or more screen positives may be taped on to the back of a sheet of drafting film (to retain face-to-face contact when copied) and the legend, description or detail added on the front. Or a diazo film intermediate may be made from the screened positive using a sheet of sufficient size to carry all additional information drawn on the reverse (matt) side of the diazo film. Or it can be trimmed and dropped into a 'window' cut in a draughting sheet. The final copies then comprise the photograph with its immediate visual clarity combined with all necessary explanations, instructions, or drawn detail where the photograph cannot for some reason give this. Photo-drawing is especially useful where electric wiring or service pipe layouts have to be seen 'in depth', which is neither easy to draw nor to read when drawn.

Production of printed forms

A printed form may be any arrangement of text and lines, usually with spaces or boxes into which the user has to insert information, either handwritten, typewritten or by machine of some kind. Thus the arrangement or layout must be exact and convenient to the purpose. Traditionally, the text would be set in metal type, and then 'rules' added to create columns, boxes, etc. The result by letterpress alone is not very elegant and may require much time to prepare.

An alternative approach may be made by phototypesetting, possibly on stripping film but not necessarily so, and assembling the text on a temporary support sheet working on a light-table over a grid or layout diagram. It is then possible to strip-in the lines from similar film, made or purchased ready for this purpose. From the whole assembly a diazo film intermediate gives a permanent one-piece film from which printing-plates or contact negatives, etc., can be made.

If the refinement of phototypesetting is not available, the repro-pull approach may be made. The text is set in normal metal type and arranged in the forme in the correct positions, but without rules. The proof is made on paper with a baryta-like surface but supplied for this purpose and with good UV transmission. The baryta surface not only accepts a good impression from the type, but also allows the lines to be added with Indian ink in a ruling-pen.

The final result can be photographed in the camera to give a negative with or without change in size; contacted to silver film for either a negative or an auto-positive; or contacted to diazo film as previously described. Auto-reversal silver film also permits an excellent same-size film positive by reflex printing from such an 'opaque' copy.

APPENDIX

Three of the processes mentioned in the Introduction – stencil and spirit duplicating, infra-red, and electro-photography – have not been treated extensively in this book. As methods of producing copies, these are excluded from consideration as photo-chemical systems because they rely on mechanisms other than those used in the photo-reproduction and photo-mechanical processes. The following notes briefly explain where their mechanisms differ from those already described.

Stencil duplicating

The simplest stencil is prepared by typing on to a waxed sheet, or by writing on it with a special stylus. The waxed sheet is made from a long-fibre paper with an open structure quite unsuitable for most ordinary purposes. Wax impregnation leaves the fibres embedded in a layer of wax and the sheet then has sufficient strength to be handled normally. The action of typing or writing is to create openings in the sheet by removing the wax in the areas which are to print. The result is very similar to the stencil of screen printing, and in use the technique is the same: ink is forced through the openings on to the surface of the copy-sheet. For duplicating purposes, the ink tended to be "thin" – that is, of low viscosity – and the paper of high absorbency so that the ink can sink into the paper quickly and avoid smudging. There are also photo-methods of preparing stencils for duplicating, using the principles already described in the main text. One attraction of this technique is simplicity and ability to use readily-available equipment in a simple fashion.

Spirit duplicating

Spirit duplicating is in a somewhat similar category, the glazed paper masters being prepared manually or by typewriting when backed up with a special carbon paper. The pressure of typing transfers the wax layer from the carbon paper to the back of the master sheet. This wax layer contains a dye or dyes which are readily soluble in alcohol (hence "spirit"). Duplication is achieved by bringing the master into contact with the copy sheet, while moist with alcohol, so that some of the dye is transferred to the copy. Clearly there is a limit to the number of good copies which can be obtained from one master: 100 to 250 is typical. An attraction is that a single master can be made in two or more different colours, the multi-colour copies then being made in a single duplicating action. It is possible to prepare the masters by using infra-red radiation by the techniques described below, but this does not place spirit duplicating in the group of photo-chemical processes with which we have been concerned.

Copying by infra-red radiation

There are various ways in which infra-red radiation may be used to produce a record or copy of an original. It has already been explained that infra-red, visible, and ultra-violet radiations are essentially similar, being electro-magnetic in nature and differing in wavelength; also that the "energy-content", or quantum of energy, associated with the visible blue regions and ultra-violet regions corresponds in size to the energy-bonds of many chemical substances: hence the ability to induce a photo-chemical change. The energy-content of visible red light is only half that of ultra-violet at the limit of transmission by glass (page 100); and of infra-red progressively less as wavelength increases. Hence there is less possibility of inducing chemical reactions by the direct absorption of infra-red radiation. It is possible to sensitize silver halide emulsions so that they can record infra-red and this is certainly a "photo"-reaction within the general context of processes included in the main text. I have taken the view, however, that infra-red photography is one of the more specialised aspects of photography and not within the scope of reprography. It mainly concerns the criminal investigator – deciphering charred documents – or the military surveyor – penetrating mist or "breaking" a camouflage.

Thermographic methods

The main use of infra-red radiation in copying relies on the fact that when the radiation is absorbed, its energy re-appears as heat and there is an increase in temperature. Thus a printed sheet exposed to infra-red radiation absorbs the radiation where the black printing-ink occurs, and only to a much less extent where the paper is bare of ink. Therefore a heat-pattern exists in the shape of the printed characters. Thermographic processes use a variety of techniques to record this heat-pattern;

231

for example, a waxy material may be brought to its melting temperature and persuaded to transfer the melted parts to a second sheet; or a chemical reaction may be triggered off in a second sheet to give a visible change of colour. In this case, the disadvantage is that the copy remains heat-sensitive and the whole sheet may darken up if allowed to become too warm; in the language used earlier, the image has not been fixed. An attraction of thermographic operation is that reflex printing is feasible, because the copy-sheet need not be affected until the radiation absorbed by the original generates the heat-pattern to which the copy-sheet responds (see page 210).

In view of this mechanism of temperature-rise where the infra-red has been absorbed, and a secondary physical or chemical step to utilise the resultant heat-pattern, thermographic methods do not involve photo-chemical reactions as do the diazo, dichromate, ferro etc. processes. A more detailed account of such methods will be found in *Perspective*, Vol. 7, No. 1, pp. 28–46 (1965).

Electrophotography

Electrophotography has been mentioned mainly in connection with its effective speed as a result of amplification of the primary image by use of toner powder (particles of pigmented resin). It is on this account the only practical photo-process comparable to silver halide materials with respect to possible use in the camera. The principle on which it functions is again physical rather than chemical, and is referred to as *photo-conduction*.

The sensitive surface of an electrophotographic material has to behave as an electrical *insulator* in the dark, and an electrical conductor when illuminated; hence, *photo*-conductor. The electrophotographic function is then to place on the surface an electrical charge ("static" electricity, i.e. not an electrical current which flows). This must be done under dark conditions. When the charged surface is illuminated, say on the base-board of an enlarger with a projected image, the insulating layer becomes conductive wherever light falls upon it. In these places, the static electrical charges can then leak away through the photo-conductor to the base layer which is permanently conductive.

Thus, exposure results in an invisible image of residual static electrical charges. To make this visible, a toner powder is brought into contact and adheres to the charge pattern but not to the discharged parts where light acted. The toner-powder image may then be transferred and *fixed* to the final copy-sheet. Reversal processing is also possible.

232

Xerography

In xerography, the photo-conductor is a selenium plate or drum. The powder image may be transferred and fixed, either by heating to fuse the powder on the final sheet; or by exposing the sheet to solvent vapour which also attaches the powder to the sheet. It can be transferred to suitable base material to give a lithographic master.

Electrofax

In the Electrofax process, the photo-conductor is a layer of resin containing zinc oxide particles. Charging, exposing and developing are similar to xerography and the image may be heat-fixed. As the zinc oxide/resin layer is normally applied to paper, this is used as the final copy without transfer of the powder image. A more usual development of the Electrofax-type print is with the toner powder suspended in a hydrocarbon liquid. The toner particles migrate from the liquid to the charged surface, and evaporation of the volatile liquid soon leaves the print dry and the toner fixed to the resin surface by slight solvent action. Electrofax-type prints can give lithographic masters for runs of 100 up to 2000 copies.

Elfasol

A third group of electrophotographic processes relies on simple layers of organic substances as the photo-conductor. These do not require the zinc oxide of the Electrofax method to give photo-conduction. The main interest at the moment in organic photo-conductors, is in the lithographic application. Metal plates suitable for lithographic work and coated with the organic layer, are charged, exposed in a camera direct to the copy, and developed with powder. This is then fused whereupon the resin of the toner powder combines with the organic substance of the layer. All remaining parts of the organic layer can then be completely dissolved away, leaving the fused image on completely bare metal. This is an excellent plate for lithography, with a very robust image capable of long runs.

The electrophotographic processes are generally sensitive to radiations in the visible region, and the zinc oxide layers can be dye-sensitized very effectively. These processes therefore utilise lower-energy radiation than the photo-chemical systems. While admittedly the process of photo-conduction involves energy transfers of one sort or another, these are essentially physical mechanisms involving changes of state, but not of chemical composition.

INDEX

236

INSTITUTE OF REPROGRAPHIC TECHNOLOGY

The Institute enables you to meet people concerned with reprographics, principally at Regional meetings, to hear lectures, exchange ideas and make educational visits. It arranges conferences, demonstrations, lectures and exhibitions. There is a syllabus of training for students of reprographic technology and courses are available at technical colleges. The Institute also publishes "REPRO" a quarterly Journal, which is issued free to members.

Further information and membership application forms can be obtained from:-

The Secretary,
INSTITUTE OF REPROGRAPHIC TECHNOLOGY,
157, Victoria Street, London, S.W.1., England.
Telephone: 01-834 3324